Richard Harwood and Ian Lodge

Cambridge IGCSE

Chemistry

Workbook

Third edition

CAMBRIDGE
UNIVERSITY PRESS

CAMBRIDGE UNIVERSITY PRESS
Cambridge, New York, Melbourne, Madrid, Cape Town,
Singapore, São Paulo, Delhi, Tokyo, Mexico City

Cambridge University Press
The Edinburgh Building, Cambridge CB2 8RU, UK

www.cambridge.org
Information on this title: www.cambridge.org/9780521181174

© Cambridge University Press 2011

Workbook first published 2011
3rd printing 2011

Printed in the United Kingdom at the University Press, Cambridge

A catalogue record for this publication is available from the British Library

ISBN 978-0-521-18117-4 Paperback

Cover image: Peeling paint on rusting steel. © Martyn F. Chillmaid / Science Photo Library

NOTICE TO TEACHERS
References to experiments contained in this publication are provided 'as is' and
information provided is on the understanding that teachers and technicians shall
undertake a thorough and appropriate risk assessment before undertaking any of
the experiments listed. Cambridge University Press makes no warranties, representations
or claims of any kind concerning the experiments. To the extent permitted by law,
Cambridge University Press will not be liable for any loss, injury, claim, liability or damage
of any kind resulting from the use of the experiments.

Contents

Introduction   iv

1 Planet Earth   1
1.1 Global warming and the 'greenhouse effect'   1
1.2 Hydrogen as a fuel   4

2 The nature of matter   6
2.1 Changing physical state   6
2.2 Chromatography at the races   8
2.3 Atomic structure   10

3 Elements and compounds   12
3.1 Periodic patterns in the properties of
     the elements   12
3.2 The chemical bonding in simple
     molecules   13
3.3 The nature of ionic lattices   15
3.4 Making magnesium oxide – a quantitative
     investigation   16

4 Chemical reactions   20
4.1 Key chemical reactions   20
4.2 The action of heat on metal carbonates   22
4.3 The nature of electrolysis   23
4.4 Displacement reactions of the halogens   25

5 Acids, bases and salts   27
5.1 Acid and base reactions – neutralisation   27
5.2 The analysis of titration results   28
5.3 Thermochemistry – investigating the
     neutralisation of an acid by an alkali   30
5.4 Deducing a formula from a precipitation
     reaction   34

6 Quantitative chemistry   35
6.1 Calculating formula masses   35
6.2 A sense of proportion in chemistry   36
6.3 Calculating the percentage of certain elements
     in a compound and empirical formulae   37
6.4 Calculations involving solutions   38
6.5 Finding the mass of 5 cm of magnesium ribbon   40

7 How far? How fast?   42
7.1 Terms of reaction   42
7.2 The collision theory of reaction rates   43
7.3 The influence of surface area on
     the rate of reaction   45
7.4 Finding the rate of a reaction producing a gas   47
7.5 Reversible reactions involving
     inter-halogen compounds   50

8 Patterns and properties of metals   52
8.1 Group1: The alkali metals   52
8.2 The reactivity series of metals   53
8.3 Energy from displacement reactions   55

9 Industrial inorganic chemistry   59
9.1 Metal alloys and their uses   59
9.2 Extracting aluminium by electrolysis   60
9.3 The importance of nitrogen   61
9.4 Making sulfuric acid industrially   65
9.5 Concrete chemistry   66

10 Organic chemistry   68
10.1 Families of hydrocarbons   68
10.2 Unsaturated hydrocarbons (the alkenes)   70
10.3 The alcohols as fuels   71
10.4 Reactions of ethanoic acid   75

11 Petrochemicals and polymers   77
11.1 Essential processes of the petrochemical
      industry   78
11.2 Addition polymerisation   79
11.3 The structure of man-made fibre molecules   80
11.4 Condensation polymerisation   81
11.5 The analysis of condensation polymers   83

12 Chemical analysis and investigation   85
12.1 Titration analysis   85
12.2 Chemical analysis   87
12.3 Experimental design   89

Introduction

This workbook contains exercises designed to help you to develop the skills you need to do well in your IGCSE Chemistry examination.

The IGCSE examination tests three different Assessment Objectives, which we call 'skills' in this workbook. These are:

Skill A   Knowledge with understanding
Skill B   Handling information and problem solving
Skill C   Experimental skills and investigations

In the examination, about 50% of the marks are for Skill A, 30% for Skill B and 20% for Skill C.

Just learning your work and remembering it is, therefore, not enough to make sure that you get the best possible grade in the exam. Half of all the marks are for Skills B and C. You need to be able to use what you have learnt and apply it in unfamiliar contexts (Skill B) and to demonstrate experimental and data handling skills (Skill C).

There are lots of exam-style questions in your coursebook which, together with the material on the accompanying CD-ROM, are aimed at helping you to develop the examination skills necessary to achieve your potential in the exams. Chapter **12** in the coursebook also deals with the experimental skills you will need to apply during your course. This workbook adds detailed exercises to help you further. There are some questions that simply involve remembering things you have been taught (Skill A), but most of the exercises require you to use what you have learned to work out, for example, what a set of data means, or to suggest how an experiment might be improved: they are aimed at developing Skills B and C.

There are a good many opportunities for you to draw graphs, read scales, interpret data and draw conclusions. These skills are heavily examined in Paper 6 of the CIE syllabus and so need continuous practice to get them right. Self-assessment check lists are provided to enable you to judge your work according to criteria similar to those used by examiners. You can try marking your own work using these. This will help you to remember the important points to think about. Your teacher should also mark the work, and will discuss with you whether your own assessments are right.

The workbook follows the same chapter breakdown as the coursebook. It is not intended that you should necessarily do the exercises in the order printed, but that you should do them as needed during your course. There are questions from all sections of the syllabus and one aim has been to give a broad range of examples of how the syllabus material is used in exam questions. The workbook is aimed at helping all students that are taking the Chemistry course. In some exercises, you will see this symbol in the margin:

S

This indicates that the exercise is intended for students who are studying the Supplement content of the syllabus as well as the Core.

We trust that the range and differing approaches of the exercises will help you develop a good understanding of the course material and the skills to do really well in the exams.

1 Planet Earth

Useful equations

carbon dioxide + water → glucose + oxygen

glucose + oxygen → carbon dioxide + water

$6CO_2 + 6H_2O \rightarrow C_6H_{12}O_6 + 6O_2$ photosynthesis

$C_6H_{12}O_6 + 6O_2 \rightarrow 6CO_2 + 6H_2O$ respiration

Exercise 1.1 Global warming and the 'greenhouse effect'

This exercise will help in developing your skills at processing unfamiliar data and making deductions from novel sources.

The diagram shows a simplified carbon cycle.

a Describe the process of photosynthesis in simple terms.

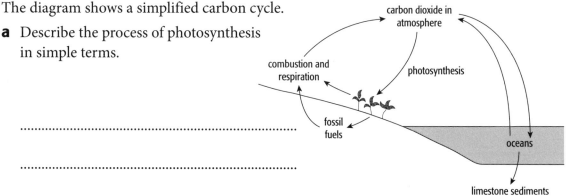

....

....

....

The '**greenhouse effect**' is caused by heat from the Sun being trapped inside the Earth's atmosphere by some of the gases which are present – their molecules absorb infrared radiation. As the amount of these 'greenhouse gases' increases, the mean (average) temperature of the Earth increases. It is estimated that if there were no 'greenhouse effect' the Earth's temperature would be cooler by 33 °C on average. Some of the gases which cause this effect are carbon dioxide, methane and oxides of nitrogen (NO_x).

Global warming: Since the burning of fossil fuels started to increase in the late 19th century, the amount of carbon dioxide in the atmosphere has increased steadily. The changes in the mean temperature of the Earth have not been quite so regular. Below are some data regarding the changes in mean temperature of the Earth and amount of carbon dioxide in the atmosphere. The first table (Table **1**) gives the changes over recent years, while the second table gives the longer term changes (Table **2**). The mean temperature is the average over all parts of the Earth's surface over a whole year. The amount of carbon dioxide is given in ppm (parts of carbon dioxide per million parts of air).

Year	CO_2 / ppm	Mean temperature / °C
1980	338	14.28
1982	340	14.08
1984	343	14.15
1986	347	14.19
1988	351	14.41
1990	354	14.48
1992	356	14.15
1994	358	14.31
1996	361	14.36
1998	366	14.70
2000	369	14.39
2002	373	14.67
2004	377	14.58
2006	381	14.63
2008	385	14.51

Table 1

Year	CO_2 / ppm	Mean temperature / °C
1880	291	13.92
1890	294	13.81
1900	297	13.95
1910	300	13.80
1920	303	13.82
1930	306	13.96
1940	309	14.14
1950	312	13.83
1960	317	13.99
1970	324	14.04
1980	338	14.28

Table 2

b Plot these results on the grid using the left-hand *y*-axis for amount of carbon dioxide and the right-hand *y*-axis for mean temperature. Draw two separate graphs to enable you to compare the trends (use graph paper if you need a larger grid).

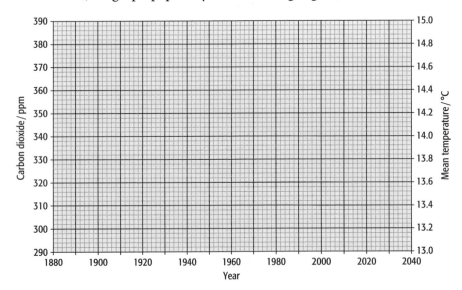

Use the check list below to give yourself a mark for your graph.
For each point, award yourself: 2 marks if you did it really well
1 mark if you made a good attempt at it, and partly succeeded
0 marks if you did not try to do it, or did not succeed.

Self-assessment check list for a graph:

Check point	Marks awarded	
	You	Your teacher
You have plotted each point precisely and correctly for both sets of data; using the different scales on the two vertical axes.		
You have used a small, neat cross or dot for the points of one graph.		
You have used a small, but different, symbol for the points of the other graph.		
You have drawn the connecting lines through one set of points accurately – using a ruler for the lines.		
You have drawn the connecting lines through the other set of points accurately – using a different colour or broken line.		
You have ignored any anomalous results when drawing the lines.		
Total (out of 12)		

10–12 Excellent.
7–9   Good.
4–6   A good start, but you need to improve quite a bit.
2–3   Poor. Try this same graph again, using a new sheet of graph paper.
1   Very poor. Read through all the criteria again, and then try the same graph again.

c What do you notice about the trend in amount of carbon dioxide?

....

....

d What do you notice about the trend in mean temperature?

....

....

e Does the graph clearly show that an increase in carbon dioxide is causing an increase in temperature?

....

....

f Estimate the amount of carbon dioxide in the atmosphere and the likely mean temperature of the Earth in the years 2020 and 2040.

....

....

g Between the 11th century and the end of the 18th century the amount of carbon dioxide in the atmosphere varied between 275 and 280 ppm. Why did it start to rise from the 19th century onwards.

....

h Other 'greenhouse gases' are present in much smaller amounts. However they are much more effective at keeping in heat than carbon dioxide. Methane (1.7 ppm) has 21 times the effect of carbon dioxide. Nitrogen oxides (0.3 ppm) have 310 times the effect of carbon dioxide.
Name a source that releases each of these gases into the atmosphere.

methane: ....

nitrogen oxides: ....

Exercise 1.2 Hydrogen as a fuel

This exercise introduces an alternative energy source and will help develop your skills at handling information regarding unfamiliar applications.

One of the first buses to use hydrogen as a fuel was operated in Erlangen, Germany, in 1996. The hydrogen was stored in thick pressurised tanks on the roof of the bus.

a Describe **two** advantages of using hydrogen as a fuel rather than gasoline (petrol).

....

....

b Suggest **one** disadvantage of using hydrogen as a fuel.

....

S It is possible to generate electrical energy from hydrogen using a fuel cell. The structure of a typical fuel cell is shown in the diagram.

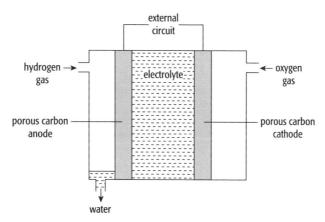

c The reaction taking place in such a fuel cell is the combustion of hydrogen. Write the overall equation for that reaction.

....

d The equation for the reaction at the anode is

$$H_2(g) + 2OH^-(aq) \rightarrow 2H_2O(l) + 2e^-$$

What type of reaction is this? Explain your answer.

....

e At the cathode oxygen molecules react with water molecules to form hydroxide ions. Write an ionic equation for this reaction.

....

2 The nature of matter

Definitions to learn

- ○ **physical state** the three states of matter are solid, liquid and gas
- ○ **condensation** the change of state from gas to liquid
- ○ **melting** the change of state from solid to liquid
- ○ **freezing** the change of state from liquid to solid at the melting point
- ○ **boiling** the change of state from liquid to gas at the boiling point of the liquid
- ○ **evaporation** the change of state from liquid to gas below the boiling point
- ○ **sublimation** the change of state directly from solid to gas (or the reverse)
- ○ **crystallisation** the formation of crystals when a saturated solution is left to cool
- ○ **filtration** the separation of a solid from a liquid using filter paper
- ○ **distillation** the separation of a liquid from a mixture using differences in boiling point
- ○ **fractional distillation** the separation of a mixture of liquids using differences in boiling point
- ○ **diffusion** the random movement of particles in a fluid (liquid or gas) leading to the complete mixing of the particles
- ○ **chromatography** the separation of a mixture of soluble (coloured) substances using paper and a solvent
- ○ **atom** the smallest part of an element that can take part in a chemical change
- ○ **proton number (atomic number)** the number of protons in the nucleus of an atom of an element
- ○ **nucleon number (mass number)** the number of protons and neutrons in the nucleus of an atom
- ○ **electron arrangement** the organisation of electrons in their different energy levels (shells)
- ○ **isotopes** atoms of the same element with different numbers of neutrons in their nuclei

Exercise 2.1 Changing physical state

This exercise will develop your understanding of the kinetic theory and the energy changes involved in changes of physical state.

The graph shows the heating curve for a pure substance. The temperature rises with time as the substance is heated.

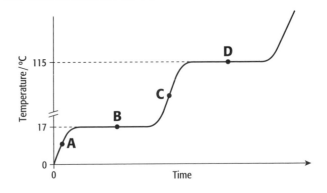

a What physical state(s) is the substance in at points **A**, **B**, **C** and **D**?

A ....   C ....

B ....   D ....

b What is the melting point of the substance? ....

c What is its boiling point? ....

d What happens to the temperature while the substance is changing state?

....

e The substance is not water. How do we know this from the graph?

....

f Complete the passage using the words given below.

| different | diffusion | gas | spread | particles |
| diffuse | random | lattice | vibrate | temperature |

The kinetic theory states that the .... in a liquid and a

.... are in constant motion. In a gas the particles are far apart from

each other and their motion is said to be .... The particles in a

solid are held in fixed positions in a regular .... In a solid the

particles can only .... about their fixed positions.

Liquids and gases are fluid states. When particles move in a fluid they can collide with

each other. When they collide they bounce off each other in ....

directions. If two gases or liquids are mixed then the different types of particle

.... out and get mixed up. This process is called ....

At the same .... particles that have a lower mass move faster than those

with higher mass. This means that the lighter particles will spread and mix more quickly; the

lighter particles are said to .... faster than the heavier particles.

g Use the data given for the substances listed below to decide which of them will be solids, liquids or gases at a room temperature of 25 °C and atmospheric pressure.

Substance	Melting point/°C	Boiling point/°C
sodium	98	883
radon	−71	−62
ethanol	−117	78
cobalt	1492	2900
nitrogen	−210	−196
propane	−188	−42
ethanoic acid	16	118

i Which substance is a liquid over the smallest range of temperature? ....

ii Which **two** substances are gaseous at −50 °C? .... and

....

iii Which substance has the lowest freezing point? ....

iv Which substance is liquid at 2500 °C? ....

v A sample of ethanoic acid was found to boil at 121 °C at atmospheric pressure. Use the information in the table to comment on this result.

....

....

Exercise 2.2 Chromatography at the races

This exercise will help you understand aspects of chromatography by considering an unfamiliar application of the technique.

Chromatography is used by the 'Horse Racing Forensic Laboratory' to test for the presence of illegal drugs in racehorses.

A concentrated sample of urine is spotted onto chromatography paper on the start line. Alongside this, known drugs are spotted. The chromatogram is run using methanol as the solvent. When finished, the paper is read by placing it under ultraviolet light. A chromatogram of urine from four racehorses is shown on the next page.

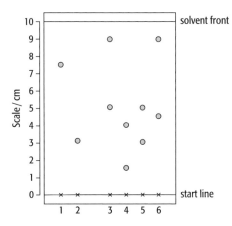

Spot	Description
1	caffeine
2	paracetamol
3	urine sample horse **A**
4	urine sample horse **B**
5	urine sample horse **C**
6	urine sample horse **D**

a State **two** factors which determine the distance a substance travels up the paper.

....

....

b From the results, the sample from one horse contains an illegal substance. State which horse, and the drug present.

....

c Give a reason for the use of this drug.

....

S **d** The results for known drugs are given as 'R_f values'.

$$R_f = \frac{\text{distance travelled by the substance}}{\text{distance travelled by the solvent}}$$

Calculate the R_f value for caffeine.

Exercise 2.3 Atomic structure

This exercise helps familiarise you with aspects of atomic structure and the uses of radioactivity.

a Choose from the words below to fill in the gaps in the passage. Words may be used once, more than once or not at all.

proton   electrons   nucleon   isotopes   protons

neutrons   nucleus   energy levels

Atoms are made up of three different particles: ....

which are positively charged; .... which have no charge;

and .... which are negatively charged. The negatively charged

particles are arranged in different .... (shells) around the

.... of the atom. The particles with a negligible mass are

the .... All atoms of the same element contain the same number of

.... and .... Atoms of the same element with

different numbers of .... are known as ....

b This part of the exercise is concerned with electron arrangements and the structure of the Periodic Table. Complete these sentences by filling in the blanks with words or numbers.

The electrons in an atom are arranged in a series of .... around the

nucleus. These shells are also called .... levels. In an atom the shell

.... to the nucleus fills first, then the next shell, and so on. There is

room for

● up to .... electrons in the first shell

● up to .... electrons in the second shell

● up to .... electrons in the third shell.

(There are 18 electrons in total when the three shells are completely full).

The elements in the Periodic Table are organised in the same way as the electrons fill the shells. Shells fill from .... to .... across the .... of the Periodic Table.

● The first shell fills up first from .... to helium.

● The second shell fills next from lithium to ....

● Eight .... go into the third shell from sodium to argon.

● Then the fourth shell starts to fill from potassium.

c In 1986, an explosion at Chernobyl in the Ukraine released a radioactive cloud containing various radioactive isotopes. Three such isotopes are shown in the table. Use your Periodic Table to answer the following questions.

Element	Nucleon (mass) number
strontium	90
iodine	131
caesium	137

i How many electrons are there in one atom of strontium-90? ....

ii How many protons are there in one atom of iodine-131? ....

iii How many neutrons are there in an atom of caesium-137? ....

The prevailing winds carried fall-out from Chernobyl towards Scandinavia. In Sweden, caesium-137 built up in lichen which is the food eaten by reindeer. This gave rise to radioactive meat.

iv If radioactive caesium was reacted with chlorine, would you expect the caesium chloride produced to be radioactive? Explain your answer.

....

v State a beneficial use in industry of a radioactive isotope.

....

vi State a medical use of a radioactive isotope.

....

3 Elements and compounds

Definitions to learn

○ **element** a substance containing only one type of atom

○ **compound** a substance made of two, or more, elements chemically combined together

○ **Periodic Table** the table in which the elements are organised in order of increasing proton number and electron arrangement

○ **Group** a vertical column of elements in the Periodic Table; elements in the same Group have similar properties

○ **Period** a horizontal row of elements in the Periodic Table

○ **valency** the number of chemical bonds an atom can make

Exercise 3.1 Periodic patterns in the properties of the elements

This exercise will help your understanding of the periodic, or repeating, patterns shown by the elements. It will also support your understanding of the structure of the Periodic Table in Groups of elements, and help you begin to predict properties within these Groups.

One physical property that shows a periodic change linked to the Periodic Table is the melting point of an element. Below is a chart of the melting points of the elements in the second and third Periods plotted against the proton number of the element.

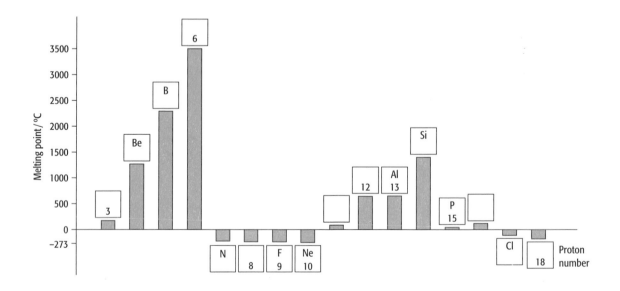

a Fill in the symbols and proton numbers missing from the boxes on the chart (7 symbols and 7 proton numbers).

b Which two elements are at the peaks of the chart? .... and

....

c To which Group do these two elements belong? ....

d The halogens are one Group of elements in the Periodic Table. Complete the following statements about the halogens by **crossing out the incorrect bold words**.

- The halogens are **metals / non-metals** and their vapours are **coloured / colourless**.
- The halogens are **toxic / non-toxic** to humans.
- Halogen molecules are each made of **one / two** atoms; they are **monatomic / diatomic**.
- Halogens react with **metal / non-metal** elements to form crystalline compounds that are salts.
- The halogens get **more / less** reactive going down the Group in the Periodic Table.
- Halogens can **colour / bleach** vegetable dyes and kill bacteria.

e Within the halogen Group there is a clear trend in physical properties. Complete the table by filling in the gaps. Use the following values when filling in the missing temperatures.
$-34\,°C$ $114\,°C$ $58\,°C$

Name of halogen	chlorine	bromine	iodine
Physical state at room temperature	....	....	....
Melting point / °C	-101	-7	....
Boiling point / °C	....	....	184

Exercise 3.2 The chemical bonding in simple molecules

This exercise will familiarise you with the structures of some simple covalent compounds and the methods we have for representing the structure and shape of their molecules.

a Many covalent compounds exist as simple molecules where the atoms are joined together with single or double bonds. A covalent bond, made up of a shared pair of electrons, is often represented by a short straight line. Complete the table overleaf by filling in the blank spaces.

Name of compound	Formula	Drawing of structure	Molecular model
hydrogen chloride	....	H—Cl	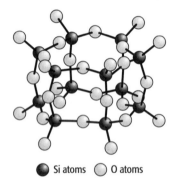
water	H_2O	O with H and H below	
ammonia	....		
....	CH_4		
ethene	....	H, H above; C=C; H, H below	
....	....	O=C=O	

S **b** Silicon dioxide is a very common compound in the crust of the Earth. It has a giant covalent structure similar to diamond. Summarise the features of the structure of silicon dioxide (silica), as shown in the diagram, by completing the following statements.

● Si atoms ○ O atoms

- The strong bonds between the atoms

  are .... bonds.

- In the crystal there are two oxygen atoms for every silicon

  atom, so the formula is ....

S • The atoms of the lattice are organised in a .... arrangement

    like diamond, with a silicon atom at the centre of each ....

• This is an example of a .... structure.

• Each oxygen atom forms .... covalent bonds.

• Each silicon atom forms .... covalent bonds.

c Graphite is one of the crystalline forms of carbon. Two of the distinctive properties of graphite are:

• it conducts electricity even though it is a non-metal, and
• it can act as a lubricant even though it has a giant covalent structure.

Give a brief explanation of these properties in the light of the structure of graphite.

i Graphite as an electrical conductor

....

....

....

ii Graphite as a lubricant

....

....

....

Exercise 3.3 The nature of ionic lattices

This exercise will help you relate the structures of ionic compounds to some of their key properties.

The diagram shows a model of the structure of sodium chloride and similar ionic crystals. The ions are arranged in a regular lattice structure – a giant ionic lattice.

On the following page are boxes containing properties of ionic compounds and their explanations. Draw lines to link each pair.

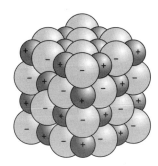

Property	Explanation
The solution of an ionic compound in water is a good conductor of electricity – such ionic substances are electrolytes.	The ions in the giant ionic structure are always arranged in the same regular way – see the diagram.
Ionic crystals have a regular shape. All the crystals of each solid ionic compound are the same shape. Whatever the size of the crystal, the angles between the faces of the crystal are always the same.	The giant ionic structure is held together by the strong attraction between the positive and negative ions. It takes a lot of energy to break down the regular arrangement of ions.
Ionic compounds have relatively high melting points.	In a molten ionic compound the positive and negative ions can move around – they can move to the electrodes when a voltage is applied.
When an ionic compound is heated above its melting point, the molten compound is a good conductor of electricity.	In a solution of an ionic compound, the positive metal ions and the negative non-metal ions can move around – they can move to the electrodes when a voltage is applied.

Exercise 3.4 Making magnesium oxide – a quantitative investigation

This exercise will develop your skills in processing and interpreting results from practical work.

Magnesium oxide is made when magnesium is burnt in air. How does the mass of magnesium oxide made depend on the mass of magnesium burnt? The practical method is described below.

Method

- Weigh an empty crucible and lid.
- Roll some magnesium ribbon round a pencil, place it in the crucible and re-weigh (not forgetting the lid).
- Place the crucible in a pipe-clay triangle sitting safely on a tripod (the lid should be on the crucible).

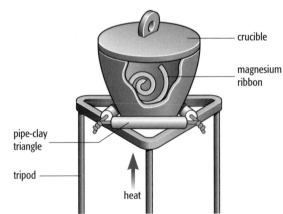

crucible

magnesium ribbon

pipe-clay triangle

tripod

heat

- Heat the crucible and contents strongly, occasionally lifting the lid to allow more air in.
- When the reaction has eased, take off the lid.
- Heat strongly for another three minutes.
- Let the crucible cool down and then weigh.
- Repeat the heating until the mass is constant.

Results

The table shows a set of class results calculated from the weights each student group obtained using this method.

Mass of magnesium/g	0.06	0.05	0.04	0.18	0.16	0.10	0.11	0.14	0.15	0.14	0.08	0.10	0.13
Mass of magnesium oxide/g	0.10	0.08	0.06	0.28	0.25	0.15	0.15	0.21	0.24	0.23	0.13	0.17	0.21

Use these results to plot a graph on the grid below relating mass of magnesium oxide made to mass of magnesium used. Remember there is one point on this graph that you can be certain of – what point is that? Include it on your graph.

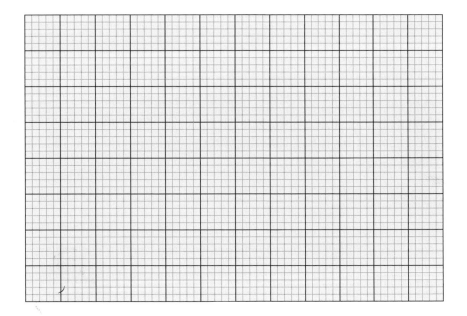

Use the check list below to give yourself a mark for your graph.
For each point, award yourself: 2 marks if you did it really well
1 mark if you made a good attempt at it, and partly succeeded
0 marks if you did not try to do it, or did not succeed.

Self-assessment check list for a graph:

Check point	Marks awarded	
	You	Your teacher
You have drawn the axes with a ruler, using most of the width and height of the grid.		
You have used a good scale for the x-axis and the y-axis, going up in 0.01s, 0.05s or 0.10s.		
You have labelled the axes correctly, giving the correct units for the scales on both axes.		
You have plotted each point precisely and correctly.		
You have used a small, neat cross for each point.		
You have drawn a single, clear best-fit line through the points – using a ruler for a straight line.		
You have ignored any anomalous results when drawing the line.		
Total (out of 14)		

12–14	Excellent.
10–11	Good.
7–9	A good start, but you need to improve quite a bit.
5–6	Poor. Try this same graph again, using a new sheet of graph paper.
1–4	Very poor. Read through all the criteria again, and then try the same graph again.

a How does the mass of magnesium oxide depend on the starting mass of magnesium?

....

b Work out from the graph the mass of magnesium oxide that you would get from 0.12 g

of magnesium (show the lines you use for this on your graph). .... g

c What mass of oxygen combines with 0.12 g of magnesium? .... g

d What mass of oxygen combines with 24 g of magnesium? .... g

S **e** What is the formula of magnesium oxide, worked out on the basis of these results? (Relative atomic masses: Mg = 24, O = 16.)

....

....

4 Chemical reactions

Definitions to learn

- ○ **synthesis** the formation of a more complex compound from its elements (or simple substances)
- ○ **decomposition** the breakdown of a compound into simpler substances
- ○ **precipitation** the sudden formation of a solid during a chemical reaction
- ○ **oxidation** the addition of oxygen to an element or compound
- ○ **reduction** the removal of oxygen from a compound
- ○ **electrolysis** the decomposition (breakdown) of an ionic compound by the passage of an electric current
- ○ **electrolyte** a compound which conducts electricity when molten or in solution in water, and is decomposed in the process
- ○ **combustion** the burning of an element or compound in air or oxygen
- ○ **displacement** a reaction in which a more reactive element displaces a less reactive element from a solution of a salt

Useful equations

$CuCO_3(s) \rightarrow CuO(s) + CO_2(g)$	thermal decomposition
$2Mg(s) + O_2(g) \rightarrow 2MgO(s)$	synthesis (oxidation)
$CuO(s) + H_2(g) \rightarrow Cu(s) + H_2O(l)$	reduction
$CH_4(g) + O_2(g) \rightarrow CO_2(g) + 2H_2O(l)$	combustion
$2KI(aq) + Cl_2(g) \rightarrow 2KCl(aq) + I_2(aq)$	displacement
$CuSO_4(aq) + Zn(s) \rightarrow ZnSO_4(aq) + Cu(s)$	displacement

Exercise 4.1 Key chemical reactions

This exercise is designed to support your understanding of the basic aspects of some important types of chemical reaction.

a Complete the diagrams to show what substances are used and what is produced in burning, respiration and rusting.

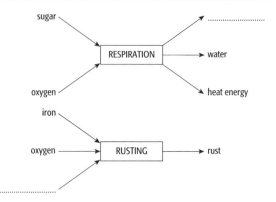

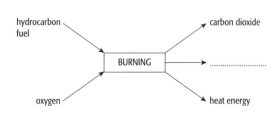

b What type of chemical change is involved in all of the above reactions? ....

c Oxidation and reduction reactions are important. There are several definitions of oxidation and reduction. Complete the following statements:

● If a substance **gains** oxygen during a reaction, it is ....

● If a substance .... oxygen during a reaction, it is **reduced.**

d The diagram shows **A** the oxidation of copper to copper(II) oxide and **B** the reduction of copper oxide back to copper using hydrogen.

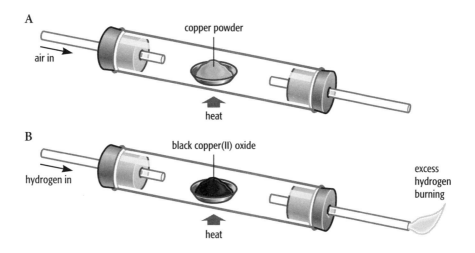

i Fill in the boxes on the equation below with the appropriate terms.

copper(II) oxide + hydrogen $\xrightarrow{\text{heat}}$ copper + water

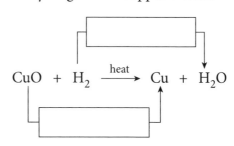

$$CuO \ + \ H_2 \ \xrightarrow{\text{heat}} \ Cu \ + \ H_2O$$

ii What type of agent is hydrogen acting as in this reaction? ....

e A further definition links oxidation and reduction to the exchange of electrons during a reaction.

REMEMBER OILRIG

i Complete the following statements.

● Oxidation is the .... of electrons.

● Reduction is the .... of electrons.

ii Fill in the boxes on the ionic equation below with the appropriate terms.

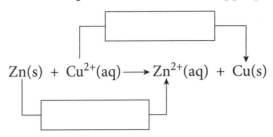

$$Zn(s) + Cu^{2+}(aq) \longrightarrow Zn^{2+}(aq) + Cu(s)$$

iii What type of agent are copper(II) ions acting as in this reaction? ....

Exercise 4.2 The action of heat on metal carbonates

This exercise will help you recall one of the major types of chemical reaction and help develop your skill at deducing conclusions from practical work.

The carbonates of many metallic elements decompose when heated.

a What type of reaction is this?

....

b Name the gas produced during the breakdown of a metal carbonate, and describe a chemical test for this gas.

....

....

c A student investigates the breakdown of five different metal carbonates using the apparatus shown.

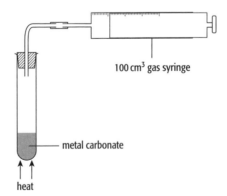

100 cm³ gas syringe

metal carbonate

heat

She heats a 0.010 mol sample of each carbonate using the blue flame of the same Bunsen burner. She measures the time it takes for 100 cm³ of gas to be collected in the gas syringe. The following table shows her results.

Carbonate	Time taken to collect 100 cm³ of gas/s
metal A carbonate	20
metal B carbonate	105
metal C carbonate	320
metal D carbonate	no gas produced after 1000 seconds
metal E carbonate	60

In fact the student used samples of calcium carbonate, copper(II) carbonate, magnesium carbonate, sodium carbonate and zinc carbonate.

Given the information that the more reactive a metal is, the less easy it is to break down the metal carbonate, complete the table to show the identity of each metal A, B, C, D and E.

Metal	Name of metal
A	....
B	....
C	....
D	....
E	....

d Write the chemical equation for the breakdown of zinc carbonate.

....

Exercise 4.3 The nature of electrolysis

This exercise will help you summarise the major aspects of electrolysis and its applications.

a Complete the following passage by using the words listed below.

anode   electrodes   current   molten   electrolyte   solution   cathode
positive   hydrogen   molecules   lose   oxygen

Changes taking place during electrolysis

During electrolysis ionic compounds are decomposed by the passage of an electric current.

For this to happen the compound must either be .... or in

.... Electrolysis can occur when an electric ....

passes through a molten .... The two rods dipping into the

electrolyte are called the .... In this situation, metals are deposited at

the .... and non-metals are formed at the ....

When the ionic compound is dissolved in water the electrolysis can be more complex.

Generally, during electrolysis .... ions move towards the

.... and negative ions move towards the ....

At the negative electrode (cathode) the metal or .... ions gain

electrons and form metal atoms or hydrogen .... At the positive

electrode (anode) certain non-metal ions .... electrons

and .... or chlorine is produced.

b Complete the passage by using the words listed below.

hydrogen	hydroxide	lower	copper	sodium	molten
cryolite	purifying	positive	concentrated		

Examples of electrolysis in industry

There are several important industrial applications of electrolysis; the most important

economically being the electrolysis of .... aluminium oxide

to produce aluminium. The aluminium oxide is mixed with molten

.... to .... the melting point of the electrolyte.

A .... aqueous solution of sodium chloride contains

...., chloride, hydrogen and .... ions. When this

solution is electrolysed, ... rather than sodium is discharged at the

negative electrode. The solution remaining is sodium hydroxide.

When a solution of copper(II) sulfate is electrolysed using ....

electrodes an unusual thing happens and the copper atoms of the

.... electrode (anode) go into solution as copper ions. At the cathode

the copper ions turn into copper atoms, and the metal is deposited on this electrode. This

can be used as a method of refining or .... impure copper.

Exercise 4.4 Displacement reactions of the halogens

> This exercise will build your understanding of a certain type of reaction and assist your skills in organising and presenting experimental observations.

The halogens – chlorine, bromine and iodine – differ in terms of their ability to displace another halogen from a solution of its salt. The following are some notes from a student's experiment. They include some rough observations from the tests carried out.

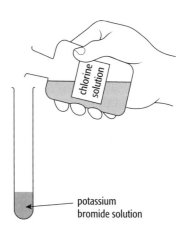

The halogens were provided as solutions in water and the test was to add the halogen to the salt solution. Solutions of potassium chloride, potassium bromide and potassium iodide were provided.

To add further observations, hexane was also available as a solvent to mix with the reaction mixture at the end of the experiment. The product was shaken with hexane and the layers allowed to separate. The colour, if any, of the hexane layer was noted.

Results

Rough notes: KCl solution with bromine or iodine solutions – no change to colourless solution – no colour in hexane layer at end.

KBr solution with iodine solution – no change to colourless solution – no colour in hexane layer at end.

KBr solution with chlorine solution – solution colourless to brown – brown colour moves to upper hexane layer at end.

KI solution with chlorine or bromine water – solution colourless to brown in both cases – purple colour in upper hexane layer at end (brown colour of aqueous layer reduced).

a Take these recorded observations and draw up a table of the results. If there is no change, then write 'no reaction'.

Use this check list to give yourself a mark for your results table.
For each point, award yourself: 2 marks if you did it really well
1 mark if you made a good attempt at it, and partly succeeded
0 marks if you did not try to do it, or did not succeed.
Self-assessment check list for results table:

Check point	Marks awarded	
	You	Your teacher
You have drawn the table with a ruler.		
The headings are appropriate and cover the observations you expect to make.		
The observations are recorded accurately, clearly and concisely – without over-elaboration.		
The table is easy for someone else to read and understand.		
Total (out of 8)		

8   Excellent.
7   Good.
5–6 A good start, but you need to improve quite a bit.
3–4 Poor. Try this same results table again, using a new sheet of paper.
1–2 Very poor. Read through all the criteria again, and then try the same results table again.

b Use the results to complete the diagram below which places the halogens tested in order of increasing reactivity.

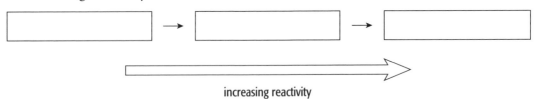

increasing reactivity

5 Acids, bases and salts

Definitions to learn

○ **acid** a substance that dissolves in water to give a solution with a pH below 7

○ **base** a substance which will neutralise an acid to give a salt and water only

○ **alkali** a base that dissolves in water

○ **pH scale** a measure of the acidity or alkalinity of a solution (scale from 0 to 14)

○ **indicator** a substance that changes colour depending on whether it is in an acid or alkali

○ **salt** an ionic substance produced from an acid by neutralisation with a base

○ **neutralisation reaction** a reaction between an acid and a base to produce a salt and water only

Useful equations

$$HCl(aq) + NaOH(aq) \rightarrow NaCl(aq) + H_2O(l)$$
$$H_2SO_4(aq) + 2KOH(aq) \rightarrow K_2SO_4(aq) + 2H_2O(l)$$
$$HNO_3(aq) + NH_3(aq) \rightarrow NH_4NO_3(aq)$$
$$CuO(s) + H_2SO_4(aq) \rightarrow CuSO_4(aq) + H_2O(l)$$

neutralisation

$$FeSO_4(aq) + 2NaOH(aq) \rightarrow Fe(OH)_2(s) + Na_2SO_4(aq)$$
$$AlCl_3(aq) + 3NaOH(aq) \rightarrow Al(OH)_3(s) + 3NaCl(aq)$$
$$AgNO_3(aq) + KI(aq) \rightarrow AgI(s) + KNO_3(aq)$$

precipitation

Exercise 5.1 Acid and base reactions – neutralisation

This exercise will help you familiarise yourself with some of the terms involved in talking about acids and bases.

Choose words from the list below to fill in the gaps in the following statements.

acid	carbon dioxide	hydrogen	hydrated	anhydrous
metal	precipitation	sodium	sulfuric	water

All salts are **ionic** compounds. Salts are produced when an alkali neutralises

an ....

In this reaction the salt is formed when a .... ion or an ammonium

ion from the alkali replaces one or more .... ions of the acid.

Salts can be crystallised from the solution produced by the neutralisation reaction. The salt

crystals formed often contain .... of crystallisation. These salts are

called .... salts. The salt crystals can be heated to drive

off the .... of crystallisation. The salt remaining is said

to be ....

Salts can be made by other reactions of acids. Magnesium sulfate can be made by reacting

magnesium carbonate with .... acid. The gas given

off is .... Water is also formed in this reaction.

All .... salts are soluble in water. Insoluble salts are usually

prepared by ....

Exercise 5.2 The analysis of titration results

This exercise will develop your understanding of some of the practical skills involved in acid–base titrations and the processing and evaluation of experimental results.

A student investigated an aqueous solution of sodium hydroxide and its reaction with hydrochloric acid. He carried out two experiments.

Experiment 1

Using a measuring cylinder, 10 cm³ of the sodium hydroxide solution was placed in a conical flask. Phenolphthalein indicator was added to the flask. A burette was filled to the 0.0 cm³ mark with hydrochloric acid (solution **P**).

The student added solution **P** slowly to the alkali in the flask until the colour just disappeared. Use the burette diagram to record the volume in the results table and then complete the column for experiment **1**.

Experiment 1 Final reading

Experiment 2

Experiment **1** was repeated using a different solution of hydrochloric acid (solution **Q**).

Use the burette diagrams to record the volumes in the results table and complete the column.

initial   final

Experiment 2

Table of results

Burette readings / cm³	Experiment 1	Experiment 2
final reading	....	....
initial reading	0.0	....
difference	....	....

a What type of chemical reaction occurs when hydrochloric acid reacts with sodium hydroxide?

....

b Write a word equation for the reaction.

....

c What was the colour change of the indicator observed?

....

d Which of the experiments used the greater volume of hydrochloric acid?

....

e Compare the volumes of acid used in experiments **1** and **2** and suggest an explanation for the difference between the volumes.

....

....

f Predict the volume of hydrochloric acid **P** that would be needed to react completely if experiment **1** was repeated with 25 cm³ of sodium hydroxide solution.

Volume of solution needed ....

Explanation

....

g Suggest **one** change the student could make to the **apparatus** used in order to obtain more accurate results.

....

Exercise 5.3 Thermochemistry – investigating the neutralisation of an acid by an alkali

This exercise introduces an unfamiliar form of titration and further develops your skills in presenting, processing and evaluating the results of practical work.

The reaction between dilute nitric acid and dilute sodium hydroxide solutions can be investigated by thermochemistry. This can be done by following the changes in temperature as one solution is added to another.

Apparatus
- polystyrene cup and beaker
- $25\,cm^3$ measuring cylinder
- $100\,cm^3$ measuring cylinder
- thermometer (0 to 100 °C)
- **safety glasses – to be used when handling the acid and alkali solutions**

Method
An experiment was carried out to measure the temperature changes during the neutralisation of sodium hydroxide solution with dilute nitric acid. Both solutions were allowed to stand in the laboratory for about 30 minutes.

$25\,cm^3$ of sodium hydroxide solution was added to a polystyrene beaker and the temperature was measured. Then $10\,cm^3$ of nitric acid was added to the alkali in the beaker and the highest temperature reached was measured. The experiment was repeated using the following volumes of acid: 20, 30, 40, 50 and $60\,cm^3$.

Results
Temperature of alkali solution at start of experiment = 21.0 °C.

The following temperatures were obtained for the different volumes of added acid used:

28.0, 35.0, 35.0, 31.0, 30.0 and 27.5 °C respectively.

a Record these results here in a suitable table.

Use this check list to give yourself a mark for your results table.
For each point, award yourself: 2 marks if you did it really well
1 mark if you made a good attempt at it, and partly succeeded
0 marks if you did not try to do it, or did not succeed.

Self-assessment check list for results table:

Check point	Marks awarded	
	You	Your teacher
You have drawn the table with a ruler.		
The headings are appropriate and have the correct units in each column / row.		
The table is easy for someone else to read and understand.		
If the table contains readings, all are to the same number of decimal places (for example 15.5, 14.2, etc).		
Total (out of 8)		

8   Excellent.
7   Good.
5–6 A good start, but you need to improve quite a bit.
3–4 Poor. Try this same results table again, using a new
       sheet of paper.
1–2 Very poor. Read through all the criteria again, and then
       try the same results table again.

b Plot a graph of the temperature of the solution against the volume of acid added to the alkali.

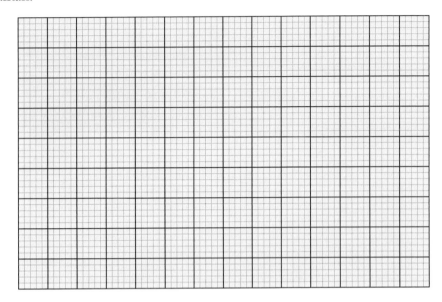

c Draw suitable lines through the points on your graph (note there are two parts to this graph so you will need to draw two straight lines through the points and extend them until they cross).

d Which point appears to be inaccurate?

....

e From these results work out the volume of acid needed to neutralise 25 cm³ of the sodium hydroxide solution. Explain why you have chosen this value.

....

....

Use the check list below to give yourself a mark for your graph.
For each point, award yourself: 2 marks if you did it really well
1 mark if you made a good attempt at it, and partly succeeded
0 marks if you did not try to do it, or did not succeed.
Self-assessment check list for a graph:

Check point	Marks awarded	
	You	Your teacher
You have drawn the axes with a ruler, using most of the width and height of the grid.		
You have used a good scale for the x-axis and the y-axis, going up in 1s, 2s, 5s or 10s.		

Check point	Marks awarded	
	You	Your teacher
You have labelled the axes correctly, giving the correct units for the scales on both axes.		
You have plotted each point precisely and correctly.		
You have used a small, neat cross for each point.		
You have drawn a single, clear best-fit line through each set of points – using a ruler for straight lines; and extended the lines to meet.		
You have ignored any anomalous results when drawing the lines.		
Total (out of 14)		

12–14 Excellent.
10–11 Good.
7–9   A good start, but you need to improve quite a bit.
5–6   Poor. Try this same graph again, using a new sheet of graph paper.
1–4   Very poor. Read through all the criteria again, and then try the same graph again.

f Why were the solutions left to stand for about 30 minutes before the experiments?

....

g Why was a polystyrene beaker used instead of a glass beaker?

....

h Suggest three improvements that would make the experiment more accurate.

....

....

....

i Write the word equation and balanced chemical equation for the reaction.

....

....

j Is the reaction exothermic or endothermic? ....

⑤ **k** The concentration of the sodium hydroxide solution is 1.0 mole per dm³. How many moles are there in 25 cm³ of this solution? (Remember there are 1000 cm³ in 1 dm³.) ....

l Look at the equation and work out how many moles of nitric acid this would react with. ....

m Calculate how many moles of acid there are in 1000 cm³ of the acid solution. What is the concentration of the acid solution in moles per dm³?

....

Exercise 5.4 Deducing a formula from a precipitation reaction

This exercise will help you familiarise yourself with an unusual method of finding the formula of an insoluble salt using precipitation.

Insoluble salts can be made using a precipitation reaction. The method can be used to find the formula of a salt. In an experiment, 6.0 cm³ of a solution of the nitrate of metal **X** was placed in a narrow test tube and 1.0 cm³ of aqueous sodium phosphate, Na_3PO_4, was added. The precipitate settled and its height was measured.

The concentration of both solutions was 1.00 mol/dm³. The experiment was repeated using different volumes of the sodium phosphate solution. The results are shown on the graph.

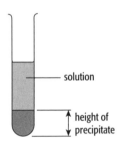

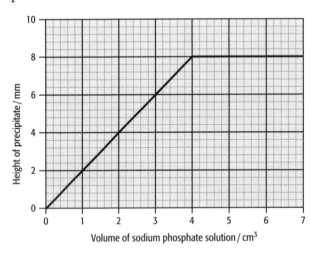

What is the formula of the phosphate of metal **X**? Give your reasoning.

....

....

....

6 Quantitative chemistry

Definitions to learn

- **relative atomic mass** the average mass of an atom on a scale where an atom of carbon-12 has a mass of 12 exactly
- **relative formula mass** the sum of all the relative atomic masses of all the atoms or ions in a compound
- **empirical formula** the formula of a compound that shows the simplest ratio of the atoms in a compound in whole numbers
- **mole** the relative formula mass of a substance in grams
- **molar gas volume** the volume occupied by one mole of any gas (24 dm³ at room temperature and pressure)

Exercise 6.1 Calculating formula masses

This exercise will develop your understanding and recall of the ideas about atomic and formula mass.

a Complete the following diagram by filling in the blanks.

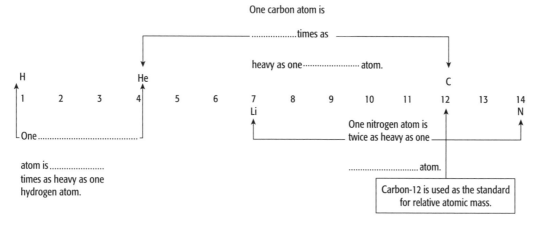

b Complete the following table of formula masses for a range of different types of substance.
(Relative atomic masses: O = 16, H = 1, C = 12, N = 14, Ca = 40, Mg = 24.)

Molecule	Chemical formula	Number of atoms or ions involved	Relative formula mass
oxygen	O_2	2 O	$2 \times 16 = 32$
carbon dioxide	....	1 C and 2 O	$1 \times 12 + 2 \times 16 =$ ....
....	H_2O	2 H and 1...........	.... = ...
ammonia	....	1 N and 3 H	.... = ...
calcium carbonate	....	$1 Ca^{2+}$ and $1 CO_3^{2-}$	.... + ... + $3 \times 16 = 100$
....	MgO	$1 Mg^{2+}$ and $1 O^{2-}$	$1 \times 24 + 1 \times 16 =$ ....
ammonium nitrate	NH_4NO_3	$1 NH_4^+$ and	$2 \times 14 +$ .... + ... $= 80$
propanol	C_3H_7OH	3 C, ... and	$3 \times 12 + 8 \times 1 +$ .... = ...

Exercise 6.2 A sense of proportion in chemistry

This exercise will familiarise you with some of the basic calculations involved in chemistry.

a Zinc metal is extracted from its oxide. In the industrial extraction process 5 tonnes of zinc oxide are needed to produce 4 tonnes of zinc. Calculate the mass of zinc, in tonnes, that is produced from 20 tonnes of zinc oxide.

b Nitrogen and hydrogen react together to form ammonia.
$$N_2 + 3H_2 \rightarrow 2NH_3$$
When the reaction is complete, 14 tonnes of nitrogen is converted into 17 tonnes of ammonia.
How much nitrogen will be needed to produce 34 tonnes of ammonia?

c The sugar lactose, $C_{12}H_{22}O_{11}$, is sometimes used in place of charcoal in fireworks. State

the total number of atoms present in a molecule of lactose. ....

d A molecule of compound **Y** contains the following atoms bonded covalently together:
- 2 atoms of carbon, (C)
- 2 atoms of oxygen, (O)
- 4 atoms of hydrogen, (H).

What is the formula of a molecule of **Y**? ....

Ⓢ Exercise 6.3 Calculating the percentage of certain elements in a compound and empirical formulae

This exercise will develop your skills in processing calculations on formula mass and empirical formulae.

a Complete the diagram to work out the formula mass of the iron oxide in the ore magnetite. (Relative atomic masses: Fe = 56, O = 16.) Then use the steps below to work out the percentage by mass of iron in this ore.

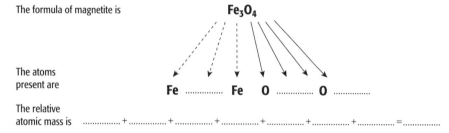

The formula of magnetite is   **Fe₃O₄**

The atoms present are   **Fe** .... **Fe   O** .... **O** ....

The relative atomic mass is .... + ... + ... + ... + ... + ... + ... = ...

- The relative formula mass of the iron oxide (Fe_3O_4) = ...

- In this formula there are .... atoms of iron, Fe.

- The relative mass of .... Fe = ...

- This means that in ....g of Fe_3O_4 there are

  .... g of iron.

- So 1 g of Fe_3O_4 contains .... g of iron.

- So 100 g of Fe_3O_4 contains .... g of iron.

- In other words, the percentage (%) of iron in Fe_3O_4 = ...%.

b Oxalic acid is an organic acid present in rhubarb and some other vegetables. The composition by mass of oxalic acid is given below, and it has an M_r of 90.

carbon = 26.7%   hydrogen = 2.2%   oxygen = 71.1%

ⓢ Calculate the empirical formula of oxalic acid.

What is the molecular formula of the acid? ....

c A volatile arsenic compound containing arsenic, carbon and hydrogen has the following composition by mass:

   arsenic = 62.5%   carbon = 30.0%   hydrogen = 7.5%

Calculate the empirical formula of this compound.

Exercise 6.4 Calculations involving solutions

This exercise will help develop your understanding of the ideas of the mole and its application to the concentration of solutions. It will develop your skills in processing practical data from titrations.

Testing the purity of citric acid

Citric acid is an organic acid which is a white solid at room temperature. It dissolves readily in water.

The purity of a sample of the acid was tested by the following method.

- Step 1: A sample of 0.48 g citric acid was dissolved in 50 cm³ of distilled water.
- Step 2: Drops of phenolphthalein indicator were added.
- Step 3: The solution was then titrated with a solution of sodium hydroxide (0.50 mol/dm³).

a i Complete the labels for the pieces of apparatus used and give the colour of the solution before titration.

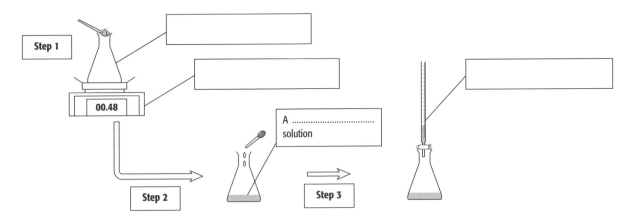

ii These were the burette readings from the titration. Complete the table by filling in the missing value (P).

Final burette reading/cm³	14.60
First burette reading/cm³	0.20
Volume of NaOH(aq) added/cm³	.... (P)

b Calculate the purity of the citric acid by following the stages outlined here.

1st stage: Calculate the number of moles of alkali solution reacted in the titration.

- P cm³ of NaOH(aq) containing 0.50 moles in 1000 cm³ were used.

- Number of moles NaOH used $= \dfrac{0.5}{1000} \times P = Q =$ .... moles.

2nd stage: Calculate the number of moles of citric acid in the sample.

- Note that 1 mole of citric acid reacts with 3 moles of sodium hydroxide.

- Then number of moles of citric acid in sample $= Q \div 3 = R =$ .... moles.

3rd stage: Calculate the mass of citric acid in the sample and, therefore, the percentage purity.

- Relative formula mass of citric acid (M_r of $C_6H_8O_7$) = ...
  (C = 12, H = 1, O = 16)
- Mass of citric acid in sample $= R \times M_r = S =$ .... g.

- Percentage purity of sample $= \dfrac{S}{0.48} \times 100 =$ ....%.

c How could the sample of citric acid be purified further?

....

....

Finding the percentage yield of hydrated copper(II) sulfate

Hydrated crystals of copper(II) sulfate-5-water were prepared by the following reactions.

$$CuO(s) + H_2SO_4(aq) \rightarrow CuSO_4(aq) + H_2O(l)$$
$$CuSO_4(aq) + 5H_2O(l) \rightarrow CuSO_4.5H_2O(s)$$

In an experiment, 25.00 cm³ of 2.0 mol/dm³ sulfuric acid was neutralised with an excess of copper(II) oxide. The yield of crystals, $CuSO_4.5H_2O$, was 7.3 g.

d Complete the following to calculate the percentage yield.

• Number of moles of H_2SO_4 in 25.00 cm³ of 2.0 mol/dm³ solution = ...

• Maximum number of moles of $CuSO_4.5H_2O$ that could be formed = ...

• Maximum mass of crystals, $CuSO_4.5H_2O$, that could be formed = ...
(The mass of one mole of $CuSO_4.5H_2O$ is 250 g.)

• Percentage yield = ...%

Exercise 6.5 Finding the mass of 5 cm of magnesium ribbon

This exercise will develop your skills in handling experimental data in novel situations.

From the chemical equation for the reaction and using the relative formula masses together with the molar volume of a gas it is possible to predict the amounts of magnesium sulfate and hydrogen that are produced when 24 g of magnesium is reacted with excess sulfuric acid.

This relationship between the mass of magnesium used and the volume of gas produced can be used to find the mass of a short piece of magnesium ribbon indirectly.

Apparatus and method

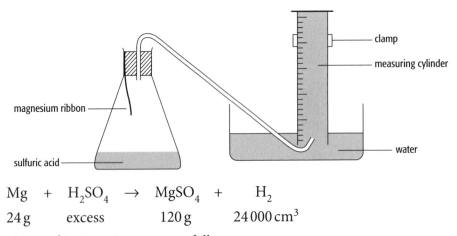

$$Mg \quad + \quad H_2SO_4 \quad \rightarrow \quad MgSO_4 \quad + \quad H_2$$
$$24\,g \quad\quad\quad excess \quad\quad\quad 120\,g \quad\quad 24\,000\,cm^3$$

The experimental instructions were as follows.

• **Wear safety goggles for eye protection.**
• Set up the apparatus as shown in the diagram with 25 cm³ of sulfuric acid in the flask.
• Make sure the measuring cylinder is completely full of water.
• Carefully measure 5 cm of magnesium ribbon and grip it below the flask stopper as shown.
• Ease the stopper up to release the ribbon and immediately replace it.
• When no further bubbles rise into the measuring cylinder record the volume of gas collected.
• Repeat the experiment twice more using 5 cm of magnesium ribbon and fresh sulfuric acid each time.
• Find the average volume of hydrogen produced.

Data handling

A student got the results shown in the table when measuring the volume of hydrogen produced.

Experiment number	Volume of hydrogen collected / cm^3
1	85
2	79
3	82
average	....

a Fill in the average of the results obtained. Can you think of possible reasons why the three results are not equal?

....

b You know that 24 g of magnesium will produce 24 000 cm^3 of hydrogen. What mass of magnesium would be needed to produce your volume of hydrogen?

....

....

This is the mass of 5 cm of magnesium ribbon. The weight is too low to weigh easily on a balance but you could weigh a longer length and use that to check your answer.

c What mass of magnesium sulfate would you expect 5 cm of magnesium ribbon to produce?

....

....

d Plan an experiment to check whether your prediction above is correct.

....

....

....

7 How far? How fast?

Definitions to learn

○ **rate of reaction** the rate of formation of the products of a chemical reaction (or the rate at which the reactants are used up)

○ **catalyst** a substance that speeds up a chemical reaction but remains unchanged at the end of the reaction

○ **enzyme** a protein that functions as a biological catalyst

○ **reversible reaction** a chemical reaction which can go both forwards and backwards; the symbol $\rightleftharpoons$ is used in the equation for the reaction

S ○ **equilibrium** a position which arises when both the forward and reverse reactions of a reversible reaction are taking place at the same speed – there is then no change in the concentration of the reactants and products unless the physical conditions are changed

Useful equations

These reactions are often used to study reaction rates or are useful examples of reversible reactions:

$$CaCO_3 + 2HCl \rightarrow CaCl_2 + H_2O + CO_2$$
$$CuSO_4.5H_2O \rightleftharpoons CuSO_4 + 5H_2O$$
$$N_2 + 3H_2 \rightleftharpoons 2NH_3$$
S
$$2H_2O_2(l) \rightarrow 2H_2O(l) + O_2(g)$$
$$Na_2S_2O_3(aq) + 2HCl(aq) \rightarrow 2NaCl(aq) + SO_2(g) + H_2O(l) + S(s)$$
$$2SO_2(g) + O_2(g) \rightleftharpoons 2SO_3(g)$$

Exercise 7.1 Terms of reaction

This exercise should help you familiarise yourself with certain key terms relating to the progress of chemical reactions.

Draw lines to match the terms on the left with the correct statement in the diagram below.

Term	Statement
chlorophyll	a substance that speeds up a chemical reaction
catalyst	the industrial process for making ammonia
Contact process	a photochemical reaction that produces glucose from carbon dioxide and water
reversible reaction	a reaction in which the products may react to produce the original reactants
photosynthesis	a reaction in which heat energy is given out to the surroundings
Haber process	the green pigment in leaves that captures energy from the Sun
exothermic reaction	the industrial process for making sulfuric acid

⑤ Exercise 7.2 The collision theory of reaction rates

This exercise should help you develop an understanding of the collision (particle) theory of reactions and how changing conditions affect the rate of various types of reaction.

Complete the following table from your understanding of the factors that affect the speed (rate) of a reaction. Several of the sections have been completed already. The finished table should then be a useful revision aid.

Factor affecting the reaction	Types of reaction affected	Change made in the condition	Effect on rate of reaction
concentration	all reactions involving solutions or reactions involving gases	an increase in the concentration of one, or both, of the .... means there are more particles in the same volume	increases the rate of reaction as the particles .... more frequently

Continued . . .

Factor affecting the reaction	Types of reaction affected	Change made in the condition	Effect on rate of reaction
pressure	reactions involving .... only	an increase in the pressure	greatly .... the rate of reaction – the effect is the same as that of an increase in ....
temperature	all reactions	an increase in temperature – this means that molecules are moving .... and collide more ....; the particles also have more .... when they collide	.... the rate of reaction
particle size	reactions involving solids and liquids, solids and gases or mixtures of solids	use the same mass of a solid but make the pieces of solid	greatly increases the rate of reaction
light	a number of photochemical reactions including photosynthesis, the reaction between methane and chlorine, and the reaction on photographic film	reaction in the presence of .... or UV light	greatly increases the rate of reaction
using a catalyst	slow reactions can be speeded up by adding a suitable catalyst	reduces amount of .... required for the reaction to take place: the catalyst is present in the same .... at the end of the reaction	.... the rate of reaction

Exercise 7.3 The influence of surface area on the rate of reaction

This exercise should help develop your skills in presenting and manipulating experimental data. You will also be asked to interpret data and draw conclusions from it.

A useful experiment that shows the effect of varying the surface area of a solid on reaction rate is based on the fact that hydrochloric acid reacts with calcium carbonate to produce the gas carbon dioxide.

The experiment was set up as shown below using identical masses of marble chips. Flask **A** contains larger pieces of marble chips and Flask **B** contains smaller pieces. The same concentration and volume of acid was used in both flasks.

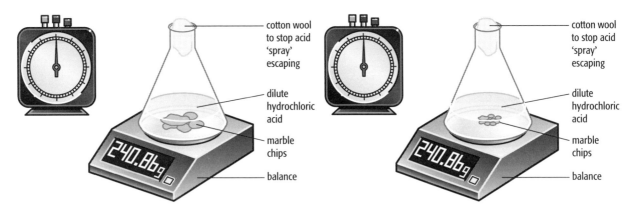

*Flask **A**: larger pieces of marble chips*   *Flask **B**: smaller pieces of marble chips*

The flasks were quickly and simultaneously set to zero on the balances. The mass loss of the flasks was then recorded over time.

a Write the word equation for the reaction between marble chips (calcium carbonate) and dilute hydrochloric acid.

....

b What causes the loss in mass from the flasks?

....

....

Readings on the digital balance were taken every 30 seconds.

For large pieces of marble chips (Flask A) , readings (in g) were:
0.00   −0.21   −0.46   −0.65   −0.76
−0.81   −0.91   −0.92   −0.96   −0.98   −0.98   −1.00   −0.99   −0.99.

For small pieces of marble chips (Flask B), readings (in g) were:
0.00   −0.51   −0.78   −0.87   −0.91
−0.94   −0.96   −0.98   −0.99   −0.99   −0.99   −1.00   −0.99   −1.00.

c Create a suitable table showing how the mass of carbon dioxide produced (equal to the loss of mass) varies with time for the two experiments.

d Plot the two graphs on the grid.

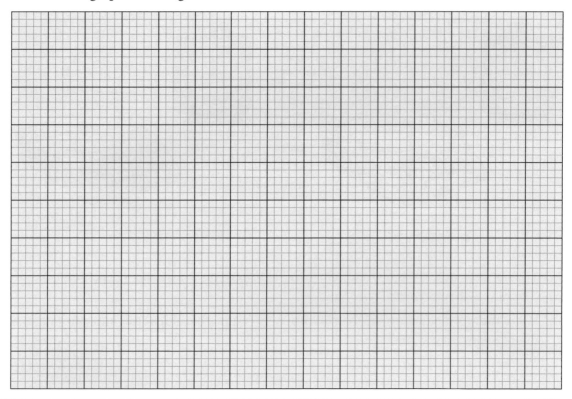

Use the check list below to give yourself a mark for your graph.
For each point, award yourself: 2 marks if you did it really well
1 mark if you made a good attempt at it, and partly succeeded
0 marks if you did not try to do it, or did not succeed.

Self-assessment check list for a graph:

Check point	Marks awarded	
	You	Your teacher
You have drawn the axes with a ruler, using most of the width and height of the grid.		
You have used a good scale for the x-axis and the y-axis, going up in 0.25s, 0.5s, 1s or 2s.		

Check point	Marks awarded	
	You	Your teacher
You have labelled the axes correctly, giving the correct units for the scales on both axes.		
You have plotted each point precisely and correctly.		
You have used a small, neat cross or dot for each point.		
You have drawn a single, clear best-fit line through each set of points.		
You have ignored any anomalous results when drawing the line through each set of points.		
Total (out of 14)		

12–14 Excellent.
10–11 Good.
7–9 A good start, but you need to improve quite a bit.
5–6 Poor. Try this same graph again, using a new sheet of graph paper.
1–4 Very poor. Read through all the criteria again, and then try the same graph again.

e Which pieces gave the faster rate of reaction ? Explain why.

....

....

....

f Explain why, for both flasks, the same amount of gas is produced in the end.

....

....

....

Exercise 7.4 Finding the rate of a reaction producing a gas

This exercise is based on an important practical technique of gas collection using a gas syringe. Following through the exercise should help develop your skills in presenting experimental data and calculating results from it. You will also be asked how the experiment could be modified to provide further data.

Hydrogen peroxide, H_2O_2, is an unstable compound that decomposes slowly at room temperature to form water and oxygen.

$$2H_2O_2(aq) \rightarrow 2H_2O(l) + O_2(g)$$

A student investigated how the rate of decomposition depends on the catalyst. She tested two catalysts: manganese(IV) oxide (experiment **1**) and copper (experiment **2**). The volume of oxygen produced by the reaction was measured at different times using the apparatus shown.

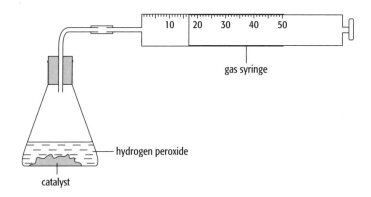

a Use the data from the diagrams below to complete the results for experiment **2** in the following table.

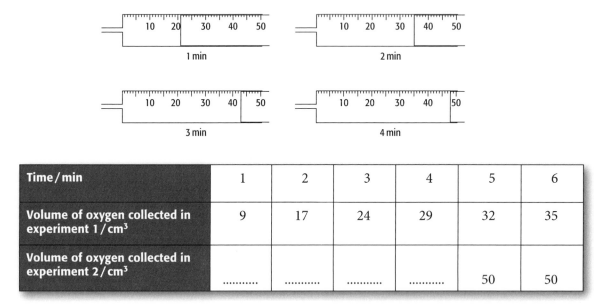

Time / min	1	2	3	4	5	6
Volume of oxygen collected in experiment 1 / cm³	9	17	24	29	32	35
Volume of oxygen collected in experiment 2 / cm³	....	....	....	....	50	50

b Plot the results from experiments **1** and **2** on the grid and draw a smooth curve through each set of points. Label the curves you draw as **exp.1** and **exp.2**.

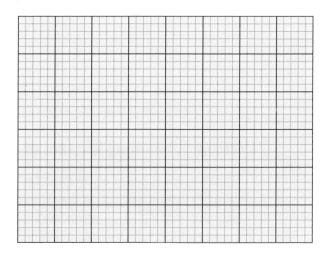

c Which of the two experiments was the first to reach completion? Explain your answer.

....

....

....

d Use your graph to estimate the time taken in experiment **1** to double the volume of oxygen produced from $15\,cm^3$ to $30\,cm^3$. Record your answers in the table, and indicate on the graph how you obtained your values.

Time taken to produce $30\,cm^3$/min	....
Time taken to produce $15\,cm^3$/min	....
Time taken to double the volume from $15\,cm^3$ to $30\,cm^3$/min	....

Experiment 1 (using manganese(IV) oxide)

e The rate (or speed) of a reaction may be calculated using the formula:

$$\text{rate of reaction} = \frac{\text{volume of oxygen produced}/cm^3}{\text{time taken}/min}$$

Using the two graphs and the above formula calculate the rate of each reaction after the first 2.5 minutes for each experiment.

f From your answer to **e**, suggest which is the better catalyst, manganese(IV) oxide or copper. Explain your answer.

....

....

g At the end of experiment **2** the copper was removed from the solution by filtration. It was dried and weighed. How would you predict this mass of copper would compare with the mass of copper added at the start of the experiment? Explain your answer.

....

....

h Suggest how the rate of decomposition in either experiment could be further increased.

....

....

....

❺ Exercise 7.5 Reversible reactions involving inter-halogen compounds

The aim of this exercise is to develop your ability to apply your knowledge to experimental situations that you will not previously have met.

The following experiment was carried out in a fume cupboard. A few crystals of iodine were placed at the bottom of a U-tube and chlorine gas passed over them. The tube got warm and a brown liquid, iodine monochloride (ICl), was formed.

a Why was the experiment carried out in a fume cupboard?

....

....

b What evidence is there that a chemical reaction took place between the iodine and chlorine?

....

....

c Write the balanced equation for the formation of iodine monochloride.

....

S The experiment was continued further as shown in the following sequence of diagrams.

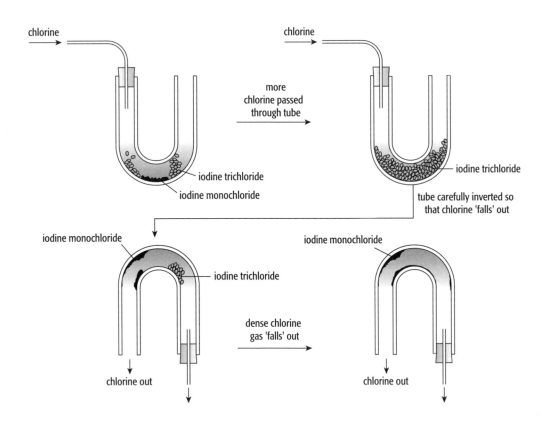

d What conclusions can be drawn from this experiment?

....

....

....

e Write balanced equations for any reactions that occur following the initial formation of iodine monochloride.

....

....

....

8 Patterns and properties of metals

Definitions to learn

- ○ **alkali metal** a reactive metal in Group I of the Periodic Table; alkali metals react with water to produce alkaline solutions
- ○ **transition metal** a metal in the central block of the Periodic Table; transition metals are hard, dense metals that form coloured compounds and can have more than one valency
- ○ **reactivity series** a listing of the metals in order of their reactivity
- **Ⓢ** ○ **electrochemical cell** a cell made up of two metal electrodes of different reactivity placed in an electrolyte; the voltage set up depends on the difference in reactivity between the two metals

Useful equations

$2Na + 2H_2O \rightarrow 2NaOH + H_2$

$Fe_2O_3 + 2Al \rightarrow Al_2O_3 + 2Fe$

Ⓢ $Zn(s) + CuSO_4(aq) \rightarrow ZnSO_4(aq) + Cu(s)$

$2KNO_3(s) \rightarrow 2KNO_2(s) + O_2(g)$

$2Pb(NO_3)_2(s) \rightarrow 2PbO(s) + 4NO_2(g) + O_2(g)$

Exercise 8.1 Group I: The alkali metals

This exercise should help you learn certain key properties of the alkali metals, and help develop the skills of predicting the properties of unfamiliar elements from the features of those that you have learnt.

Caesium is an alkali metal. It is in Group I of the Periodic Table.

a State **two** physical properties of caesium.

....

....

b State the number of electrons in the outer shell of a caesium atom. ....

c Complete the table on the next page to estimate the boiling point and atomic radius of caesium. Comment also on the reactivity of potassium and caesium with water.

Group I metal	Density/ g/cm^3	Radius of metal atom/nm	Boiling point/°C	Reactivity with water
sodium	0.97	0.191	883	floats and fizzes quickly on the surface, disappears gradually and does not burst into flame
potassium	0.86	0.235	760	....
rubidium	1.53	0.250	686	reacts instantaneously, fizzes and bursts into flame then spits violently and may explode
caesium	1.88	....	....	....

d Write the word equation for the reaction of caesium with water.

....

Exercise 8.2 The reactivity series of metals

This exercise should help you familiarise yourself with certain aspects of the reactivity series. It should also help develop your skills in interpreting practical observations, and of predicting the properties of unfamiliar elements from the features of those that you have learnt.

Using the results of various different types of chemical reaction the metals can be arranged into the reactivity series.

a Magnesium reacts very slowly indeed with cold water but it does react strongly with steam to give magnesium oxide and a gas. Write the word equation for the reaction between magnesium and steam.

....

b Choose **one** metal from the reactivity series that will not react with steam.

....

c Choose **one** metal from the reactivity series that will safely react with dilute sulfuric acid.

....

d In each of the experiments below a piece of metal is placed in a solution of a metal salt. Complete the table of observations.

		zinc in tin(II) chloride solution	zinc in copper(II) sulfate solution	tin in copper(II) sulfate solution	silver in copper(II) sulfate solution	copper in silver nitrate solution
At start	colour of metal	grey	....	silver-coloured	silver-coloured	....
	colour of solution	colourless	....	blue	blue	colourless
At finish	colour of metal	coated with silver-coloured crystals	....	coated with brown solid	silver-coloured	coated with silver-coloured crystals
	colour of solution	colourless	....	colourless	blue	....

e Use these results to place the metals **copper**, **silver**, **tin** and **zinc** in order of reactivity (putting the most reactive metal first).

....

The reactivity series of metals on the right contains both familiar and unfamiliar elements. The unfamiliar elements are marked (*) and their common oxidation states are given. Choose metal(s) from this list to answer the following questions.

barium*	Ba (+2)
lanthanum*	La (+3)
aluminium	
zinc	
chromium*	Cr (+2),(+3),(+6)
iron	
copper	
palladium*	Pd (+2)

(S) f Which two metals would not react with dilute hydrochloric acid?

....

g Which two unfamiliar metals would react with cold water?

....

h Name an unfamiliar metal that could not be extracted from its oxide by reduction with carbon.

....

i Why should you be able to predict that metals such as iron and chromium have more than one oxidation state?

....

Exercise 8.3 Energy from displacement reactions

This exercise will help you practise the presentation and interpretation of practical experiments.

When a metal is added to a solution of the salt of a less reactive metal a displacement reaction takes place. Two examples are:

$$Fe(s) + CuSO_4(aq) \rightarrow Cu(s) + FeSO_4(aq)$$

zinc + copper sulfate → copper + zinc sulfate

The energy change involved in these reactions can be measured by adding 5 g of metal powder to 50 cm^3 of 0.5 mol/dm^3 copper (II) sulfate solution in a polystyrene cup. The temperature of the solution is taken before adding the metal. The powder is then added, the reaction mixture is stirred continuously, and temperatures are taken every 30 seconds for 3 minutes. A student took the readings below when carrying out this experiment.

Time / min	0.0	0.5	1.0	1.5	2.0	2.5	3.0
Experiment 1 (zinc): temperature / °C	21	48	62	71	75	72	70
Experiment 2 (iron): temperature / °C	21	25	32	38	41	43	44

a Plot two graphs on the grid provided and label each with the name of the metal.

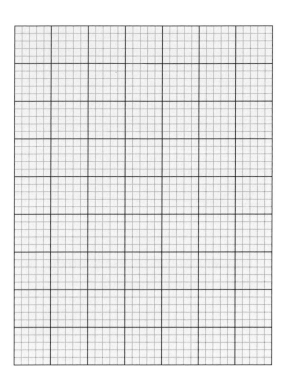

b Write the word equation for the first reaction, and the balanced symbol equation for the second.

....

....

c Which metal, iron or zinc, produced the larger temperature rise?

....

d Suggest why this metal gave the larger temperature rise.

....

....

e Comment on whether this experiment is a 'fair test'. Explain your answer.

....

....

....

Use the check list below to give yourself a mark for your graph.
For each point, award yourself: 2 marks if you did it really well
1 mark if you made a good attempt at it, and partly succeeded
0 marks if you did not try to do it, or did not succeed.
Self-assessment check list for a graph:

Check point	Marks awarded	
	You	Your teacher
You have drawn the axes with a ruler, using most of the width and height of the grid.		
You have used a good scale for the x-axis and the y-axis, going up in 1s, 2s, 5s or 10s.		
You have labelled the axes correctly, giving the correct units for the scales on both axes.		
You have plotted each point precisely and correctly.		
You have used a small, neat dot or cross for each point.		
You have drawn a single, clear best-fit line through each set of points – using a ruler for a straight line.		
You have ignored any anomalous results when drawing the lines through each set of results.		
Total (out of 14)		

12–14   Excellent.
10–11   Good.
7–9   A good start, but you need to improve quite a bit.
5–6   Poor. Try this same graph again, using a new sheet of graph paper.
1–4   Very poor. Read through all the criteria again, and then try the same graph again.

⑤ Generating electrical energy

In these metal displacement reactions the atoms of the reactive metal lose electrons to become ions. For example:

$$Zn(s) \rightarrow Zn^{2+}(aq) + 2e^-$$

f Is this reduction or oxidation? ....

The electrons produced are given to the metal ion in solution to form an atom of the displaced metal.

$$Cu^{2+}(aq) + 2e^- \rightarrow Cu(s)$$

S In an electrochemical cell these electrons are sent through a circuit to produce an electrical current. The flow of electrons is from the more reactive metal (zinc) to the less reactive metal (copper).

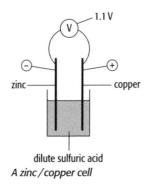

A zinc/copper cell

The voltage produced is a measure of the difference in reactivity of the two metals. A copper/iron cell produces a different voltage.

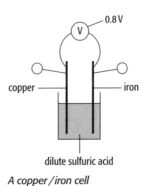

A copper/iron cell

A zinc/iron cell

g Calculate the voltage given by the zinc/iron cell.

....

h Work out which is the positive and negative electrode of the copper/iron and zinc/iron cells.

....

....

i How could an even higher voltage be obtained?

....

j Write equations to show what happens in the zinc/iron cell.

....

....

9 Industrial inorganic chemistry

Definitions to learn

- ○ **chemical plant** the reaction vessels and equipment for manufacturing chemicals
- ○ **feedstock** starting materials for chemical industrial processes
- ○ **brine** a concentrated solution of sodium chloride
- ○ **Haber process** the industrial process for the manufacture of ammonia
- ○ **Contact process** the industrial process for the manufacture of sulfuric acid

Useful equations

$Fe_2O_3 + 3CO \rightarrow 2Fe + 3CO_2$	blast furnace reaction
$CaCO_3 \rightarrow CaO + CO_2$	lime kiln reaction
$Al^{3+}(l) + 3e^- \rightarrow 3Al(l)$	extraction of aluminium
$N_2(g) + 3H_2(g) \rightleftharpoons 2NH_3(g)$	Haber process
$2SO_2(g) + O_2(g) \rightleftharpoons 2SO_3(g)$	Contact process

Exercise 9.1 Metal alloys and their uses

This exercise should help you recall details of different alloys and the basis of their usefulness.

Complete the following table on the composition and usefulness of some alloys by filling in the gaps.

Alloy	Composition	Use	Useful property
mild steel	iron: > 99.75% carbon: < 0.25%	....	....
stainless steel	iron: 74% .... : 18% nickel: 8%	.... , surgical instruments, chemical vessels for industry	....
brass	copper: 70% .... : 30%	.... instruments, ornaments	'gold' colour, harder than copper

Continued . . .

Alloy	Composition	Use	Useful property
bronze	copper: 95% ....: 5%	statues, church bells	hard, does not ....
aerospace aluminium	aluminium: 90.25% zinc: 6% magnesium: 2.5% copper: 1.25%	aircraft construction	....
solder	tin: 60% lead: 40%	....	low melting point
tungsten steel	iron: 95% tungsten: 5%	cutting edges of drill bits	....

⑨ Exercise 9.2 Extracting aluminium by electrolysis

This exercise should help you recall and understand the details of the method for extracting aluminium.

Because of its strong reactivity, aluminium must be extracted by electrolysis. The electrolyte is aluminium oxide dissolved in molten cryolite. Hydrated aluminium oxide is heated to produce the pure aluminium oxide used.

$$Al_2O_3.3H_2O \quad \rightarrow \quad Al_2O_3 + 3H_2O$$

hydrated aluminium oxide

a What type of reaction is this? Put a ring around the correct answer.

   decomposition   **neutralisation**   **oxidation**   **reduction**

b Why must the electrolyte be molten for electrolysis to occur?

   ....

c What is the purpose of the cryolite?

   ....

d In the diagram of the electrolysis cell, which letter (**A**, **B**, **C** or **D**) represents the

   cathode? ....

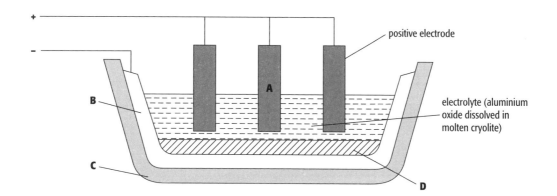

e State the name of the products formed at the anode and cathode during this electrolysis.

At anode: .... At cathode: ....

f Why do the anodes have to be renewed periodically?

....

g Complete the equation for the formation of aluminium from aluminium ions.

$$Al^{3+} + ... e^- \rightarrow Al$$

h State one use of aluminium.

....

Exercise 9.3 The importance of nitrogen

The following exercise connects the ideas surrounding the importance of nitrogen to agriculture and develops your understanding of chemical equilibria. It also develops your skills in processing and interpreting experimental results.

A simplified diagram of the nitrogen cycle is shown. Although certain bacteria in the soil convert nitrogen gas into nitrates, other bacteria convert nitrogen into ammonium salts. The ionic equation for this second reaction is

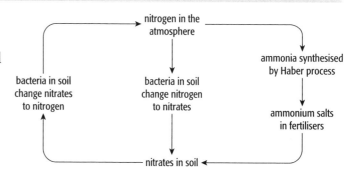

$$N_2 + 8H^+ + 6e^- \rightarrow 2NH_4^+$$

a Explain why this is a reduction reaction.

....

b In the presence of hydrogen ions, bacteria of a different type convert nitrate ions into nitrogen gas and water. Give the ionic equation for this reaction.

....

Ammonia is made by the Haber process using an iron catalyst.

$$N_2 + 3H_2 \rightleftharpoons 2NH_3 \quad \text{(the forward reaction is exothermic)}$$

The raw materials for the Haber process can be obtained from the air and from natural gas.

c Which method is used to separate pure nitrogen from other gases in the air?

....

d Describe how hydrogen can be made from hydrocarbons.

....

....

e State the essential conditions of temperature and pressure used for the Haber process.

....

f Sketch an energy profile diagram to show both the catalysed and the uncatalysed reaction. Label the diagram to show the following key features: the reactants and products, the enthalpy change for the reaction, and the catalysed and uncatalysed reactions.

The table shows how the percentage of ammonia in the mixture leaving the reaction vessel varies under different conditions.

Pressure / atm	100	200	300	400
% of ammonia at 300 °C	45	65	72	78
% of ammonia at 500 °C	9	18	25	31

5 g Use the grid to plot graphs of the percentage of ammonia against pressure at both 300 °C and 500 °C.

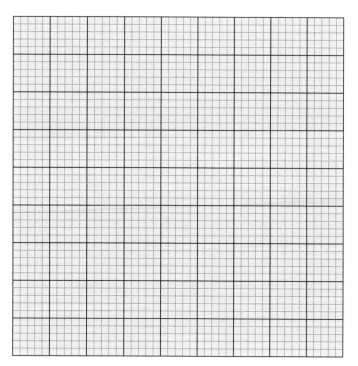

Use the check list below to give yourself a mark for your graph.

For each point, award yourself: 2 marks if you did it really well

1 mark if you made a good attempt at it, and partly succeeded

0 marks if you did not try to do it, or did not succeed.

Self-assessment check list for a graph:

Check point	Marks awarded	
	You	Your teacher
You have drawn the axes with a ruler, using most of the width and height of the grid.		
You have used a good scale for the *x*-axis and the *y*-axis, going up in useful proportions.		
You have labelled the axes correctly, giving the correct units for the scales on both axes.		
You have plotted each point precisely and correctly.		
You have used a small, neat dot or cross for each point.		

Check point	Marks awarded	
	You	Your teacher
You have drawn a single, clear best-fit line through each set of points – using a ruler for any straight line.		
You have ignored any anomalous results when drawing the line.		
Total (out of 14)		

12–14 Excellent.
10–11 Good.
7–9   A good start, but you need to improve quite a bit.
5–6   Poor. Try this same graph again, using a new sheet of graph paper.
1–4   Very poor. Read through all the criteria again, and then try the same graph again.

h What is the percentage of ammonia formed at 250 atmospheres and 300 °C?

....

i Use your graphs to estimate the percentage of ammonia formed at 400 °C and 250 atmospheres.

....

j The advantage of using a low temperature is the large percentage of ammonia formed. What is the disadvantage of using a low temperature?

....

k Suggest **two** advantages of using high pressure in the manufacture of ammonia.

....

....

The most important use of ammonia is in fertiliser production. Fertilisers are added to the soil to improve crop yields. A farmer has the choice of two fertilisers, ammonium nitrate, NH_4NO_3, or diammonium hydrogen phosphate, $(NH_4)_2HPO_4$.

l Show by calculation which of these fertilisers contains the greater percentage of nitrogen by mass.

⑤ m State one major problem caused when the nitrates from fertilisers leach from the soil into streams and rivers.

....

Exercise 9.4 Making sulfuric acid industrially

This exercise helps your understanding of chemical equilibria; and particularly the factors involved in the Contact process.

The diagram shows the three different stages in the manufacture of sulfuric acid.

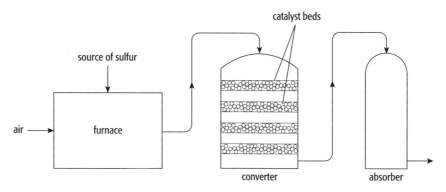

One possible source of sulfur is an ore containing zinc sulfide, ZnS. In the furnace, this sulfide ore is heated in oxygen to make zinc oxide, ZnO, and sulfur dioxide.

a Write an equation for this reaction.

....

In the converter, sulfur dioxide and oxygen are passed over a series of catalyst beds at a temperature of about 420 °C.

$$2SO_2(g) + O_2(g) \rightleftharpoons 2SO_3(g) \qquad \Delta H = -196\,kJ$$

b An increase in pressure increases the yield of sulfur trioxide. Explain the reason for this effect.

....

....

c Even though an increase in pressure increases the yield of sulfur trioxide, the reaction in the converter is carried out at atmospheric pressure. Suggest a reason for this.

....

....

d In some sulfuric acid plants, the gases are cooled when they pass from one catalyst bed to the next. Use the information given on the nature of the reaction to explain why the gases need to be cooled.

....

....

Exercise 9.5 Concrete chemistry

This exercise will aid your recall of the important uses of limestone and help your familiarity with questions asked in an unusual context.

Limestone is an important mineral resource. One use is in the making of cement. Cement is made by heating clay with crushed limestone. During this process, the calcium carbonate is first converted to calcium oxide.

$$CaCO_3 \rightarrow CaO + CO_2$$

a What name is given to this type of chemical reaction?

....

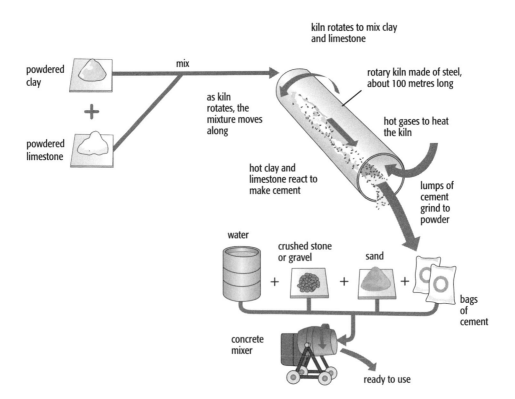

5 Concrete is then made from cement, sand and water. When it has set, concrete is slightly porous. Rain water can soak into concrete and some of the unreacted calcium oxide present dissolves to form calcium hydroxide.

b Write an equation for this reaction.

....

The aqueous calcium hydroxide in wet concrete is able to react with carbon dioxide in the air.

$$Ca(OH)_2 + CO_2 \rightarrow CaCO_3 + H_2O$$

The diagram shows how the pH can vary at different points inside a cracked concrete beam.

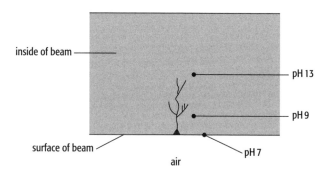

c Describe the change in pH from the surface to the centre of the beam, and explain why this variation occurs.

....

....

10 Organic chemistry

Definitions to learn

- ○ **hydrocarbon** a compound that contains carbon and hydrogen only
- ○ **saturated hydrocarbon** a hydrocarbon that contains only single covalent bonds between the carbon atoms
- ○ **alkane** a saturated hydrocarbon that contains only single covalent bonds between the carbon atoms of the chain; the simplest alkane is methane, CH_4
- ○ **alkene** an unsaturated hydrocarbon that contains at least one double bond between two of the carbon atoms in the chain; the simplest alkene is ethene, C_2H_4
- ○ **homologous series** a family of organic compounds with similar chemical properties as they contain the same functional group; alkenes, alcohols, for instance
- ○ **isomers** molecules with the same molecular formula but different structural formulae
- ○ **substitution reaction** a reaction in which one or more hydrogen atoms in a hydrocarbon are replaced by atoms of another element
- ○ **addition reaction** a reaction in which atoms, or groups, are added across a carbon–carbon double bond in an unsaturated molecule such as an alkene

Useful equations

$CH_4 + 2O_2 \rightarrow CO_2 + 2H_2O$	burning methane
$C_2H_5OH + 3O_2 \rightarrow 2CO_2 + 3H_2O$	burning ethanol
$C_6H_{12}O_6 \rightarrow 2C_2H_5OH + 2CO_2$	fermentation
$CH_4 + Cl_2 \rightarrow CH_3Cl + HCl$	substitution
$C_2H_5OH + 2[O] \rightarrow CH_3COOH + H_2O$	oxidation
$C_2H_5OH + CH_3COOH \rightleftharpoons CH_3COOC_2H_5 + H_2O$	esterification

Exercise 10.1 Families of hydrocarbons

This exercise helps you revise the key features of the families of hydrocarbons and develops your understanding of the structures of organic compounds.

a Complete the passage using only words from the list.

bromine   alkanes   hydrogen   double   chlorine   chains   petroleum

methane   ethene   ethane   colourless   propane   alkenes

The chief source of organic compounds is the naturally occurring mixture of hydrocarbons known as .... Hydrocarbons are compounds that contain carbon and .... only. There are many hydrocarbons because of the ability of carbon atoms to join together to form long ....

There is a series of hydrocarbons with just single covalent bonds between the carbon atoms in the molecule. These are saturated hydrocarbons, and they are called .... The simplest of these saturated hydrocarbons has the formula CH_4, and is called .... Unsaturated hydrocarbons can also occur. These molecules contain at least one carbon–carbon .... bond. These compounds belong to the ...., a second series of hydrocarbons. The simplest of this 'family' of unsaturated hydrocarbons has the formula C_2H_4, and is known as ....

The test for an unsaturated hydrocarbon is to add the sample to .... water. It changes colour from orange/brown to .... if the hydrocarbon is unsaturated.

b The table shows the names, formulae and boiling points of the first members of the homologous series of unsaturated hydrocarbons. Complete the table by filling in the spaces.

c Deduce the molecular formula of the alkene which has a relative molecular mass of 168.

....

Name	Formula	Boiling point/°C
....	C_2H_4	−102
propene	C_3H_6	−48
butene	C_4H_8	−7
pentene	C_5H_{10}	30
hexene	....	....

Exercise 10.2 Unsaturated hydrocarbons (the alkenes)

This exercise develops your understanding of unsaturated hydrocarbons using an unfamiliar example.

Limonene is a colourless unsaturated hydrocarbon found in oranges and lemons. The structure of limonene is shown here.

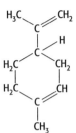

a On the structure, draw a circle around the bonds which make limonene an unsaturated hydrocarbon.

b What is the molecular formula of limonene?

....

c Describe the colour change which occurs when excess limonene is added to a few drops of bromine water.

....

The diagram shows how limonene can be extracted from lemon peel by steam distillation.

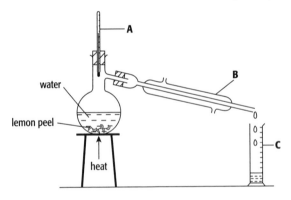

d State the name of the pieces of apparatus labelled A, B and C.

A .... B .... C ....

When limonene undergoes incomplete combustion, carbon monoxide is formed.

e What do you understand by the term **incomplete combustion**?

....

....

f State an adverse effect of carbon monoxide on health.

....

....

g The structures of some compounds found in plants are shown below.

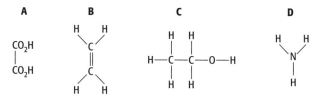

i Which one of these compounds is a carboxylic acid? ....

ii Which one of these compounds is produced by the fermentation

of glucose? ....

iii Which one of these compounds is a hydrocarbon? ....

h All hydrocarbons are covalently bonded whether saturated or unsaturated. Draw 'dot-and-cross' diagrams for methane and ethane illustrating the arrangement of the bonding electrons. You only need to draw the outer electrons of the carbon atoms.

Exercise 10.3 The alcohols as fuels

The following exercise uses information relating to the alcohols to develop your understanding of these compounds and to enhance your presentation, analysis and interpretation of experimental data concerning their property as fuels.

The table shows the formula of the first three members of the alcohol homologous series.

Alcohol	Formula
methanol	CH_3OH
ethanol	C_2H_5OH
propanol	C_3H_7OH

a Use the information given to deduce the general formula for the alcohol homologous series.

....

Ethanol, the most significant of the alcohols, can be manufactured from either ethene or glucose.

b Write an equation for the industrial production of ethanol from ethene and state the conditions under which the reaction takes place.

....

....

The fermentation (anaerobic respiration) of glucose by yeast can be represented by the following equation. The reaction is catalysed by the enzyme, zymase. After a few days the reaction stops. It has produced a 12% aqueous solution of ethanol.

$$C_6H_{12}O_6 \rightarrow 2C_2H_5OH + 2CO_2$$

c Sketch a labelled diagram to show how fermentation can be carried out.

d Suggest a reason why the reaction stops after a few days.

....

....

e Why is it essential that there is no oxygen in the reaction vessel?

....

....

f Name the products of the complete combustion of ethanol.

....

g Explain why ethanol made from ethene is a non-renewable fuel, but that made from glucose is a renewable fuel.

....

....

....

A student used this apparatus to investigate the amount of heat produced when ethanol was burnt.

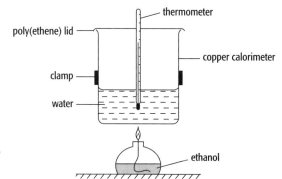

h Draw the structure of ethanol showing all atoms and bonds.

i Complete the equation for the complete combustion of ethanol.

$$C_2H_5OH + 3O_2 \rightarrow \text{..........}CO_2 + \text{..........}H_2O$$

j When 2.3 g of ethanol is burnt, 2.7 g of water is formed. Calculate the mass of water formed when 13.8 g of ethanol is burnt.

The experiment was later adapted to compare the heat released by burning four different alcohols. Each burner in turn was weighed and then the alcohol was allowed to burn until the temperature of the water had risen by 15 °C. The flame was then extinguished and the burner reweighed. The following results were obtained.

Alcohol	Formula	Mass of alcohol burnt/g
methanol	CH_3OH	0.90
ethanol	C_2H_5OH	0.70
propan-1-ol	C_3H_7OH	0.62
pentan-1-ol	$C_5H_{11}OH$	0.57

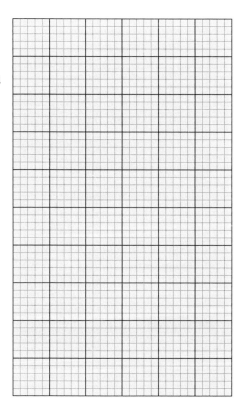

k Plot a graph showing how the mass of alcohol required varies with the number of carbon atoms in the alcohol used. Draw a smooth curve through the points.

Use the check list below to give yourself a mark for your graph.
For each point, award yourself: 2 marks if you did it really well
1 mark if you made a good attempt at it, and partly succeeded
0 marks if you did not try to do it, or did not succeed.

Self-assessment check list for a graph:

Check point	Marks awarded	
	You	Your teacher
You have drawn the axes with a ruler, using most of the width and height of the grid.		
You have used a good scale for the x-axis and the y-axis, going up in useful proportions.		
You have labelled the axes correctly, giving the correct units for the scales on both axes.		
You have plotted each point precisely and correctly.		
You have used a small, neat cross or dot for each point.		
You have drawn a single, clear best-fit line through the points – using a ruler for a straight line.		
You have ignored any anomalous results when drawing the line.		
Total (out of 14)		

12–14 Excellent.

10–11 Good.

7–9 A good start, but you need to improve quite a bit.

5–6 Poor. Try this same graph again, using a new sheet of graph paper.

1–4 Very poor. Read through all the criteria again, and then try the same graph again.

l Predict the mass of butanol, C_4H_9OH, which, on combustion, would raise the temperature of the water by 15 °C.

....

m Suggest a reason why the same temperature rise (15 °C) was used in each experiment.

....

....

(S) n One student found a different value of 0.66 g for the mass of propanol burnt in this experiment. This student had accidentally used a different isomer of propanol. Give the name and structure of this isomer.

Name ....

Structure

Exercise 10.4 Reactions of ethanoic acid

This exercise revises some aspects of the chemistry of carboxylic acids.

Acidified potassium dichromate(VI) was used to oxidise ethanol to ethanoic acid using the apparatus in the diagram.

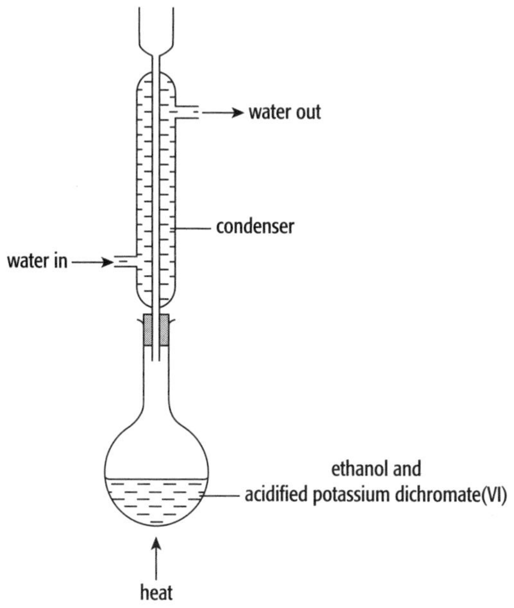

a Which of the following best describes the purpose of the condenser set up as shown? Put a ring around the correct answer.

A to prevent the conversion of ethanoic acid back to ethanol

B to prevent condensation of the oxidising agent

C to prevent escape of the unreacted alcohol

D to prevent reaction between ethanol and ethanoic acid

b Ethanoic acid, CH_3CO_2H, is a weak acid. Explain what is meant by the term **weak acid**.

....

....

S **c** Ethanoic acid reacts with sodium carbonate. Write the equation for this reaction.

....

Ethanoic acid also reacts with magnesium to form magnesium ethanoate and hydrogen.

$$Mg + 2CH_3CO_2H \rightarrow (CH_3CO_2)_2Mg + H_2$$

A student added 4.80 g of magnesium to 25.0 g of ethanoic acid. (Note the relative atomic masses H = 1, C = 12, O = 16 and Mg = 24; and that the molar volume of a gas = 24 dm^3 at r.t.p.)

d Which of the reactants, magnesium or ethanoic acid, is used in excess in this experiment? Explain your answer.

....

....

....

....

e Calculate both the number of moles of hydrogen and the volume of hydrogen that would be formed at r.t.p.

11 Petrochemicals and polymers

Definitions to learn

○ **fossil fuel** a fuel formed underground from previously living material by the action of heat and pressure over geological periods of time

○ **cracking** a thermal decomposition reaction in which a long-chain saturated alkane is broken down to a shorter alkane, usually with the formation of an alkene

○ **catalytic cracking** cracking carried out in the presence of a catalyst

○ **monomer** the small molecules from which polymers are built by joining them together

○ **polymer** a long-chain molecule made by joining many monomer molecules together

○ **polymerisation** the process by which a long-chain polymer is made from its monomers

○ **addition polymerisation** a polymerisation process in which the monomers contain a carbon–carbon double bond and polymerisation takes place by addition reactions

○ **condensation polymerisation** a polymerisation process in which the linking of the monomers takes place by a condensation reaction in which a small molecule, usually water, is eliminated

Useful equations

$$C_{10}H_{22} \rightarrow C_8H_{18} + C_2H_4 \qquad \text{cracking}$$

$$\left. \begin{array}{l} nC_2H_4 \rightarrow -(C_2H_4)_n- \\ nCH_2CHCl \rightarrow -(CH_2CHCl)_n- \end{array} \right\} \text{ addition polymerisation}$$

The following equations represent the condensation reactions to form nylon, a protein, a polyester and starch respectively. Remember you need only to show how the linkage is formed in each case.

- nylon

- protein

- polyester

- starch

Exercise 11.1 Essential processes of the petrochemical industry

This exercise aids you in recalling and understanding two of the main processes of the petrochemical industry.

Petroleum (crude oil) is a raw material which is processed in an oil refinery. Two of the processes used are **fractional distillation** and **cracking**.

a The diagram shows the fractional distillation of petroleum. Give the name and a major use for each fraction.

A: ....

B: ....

C: ....

D: ....

E: ....

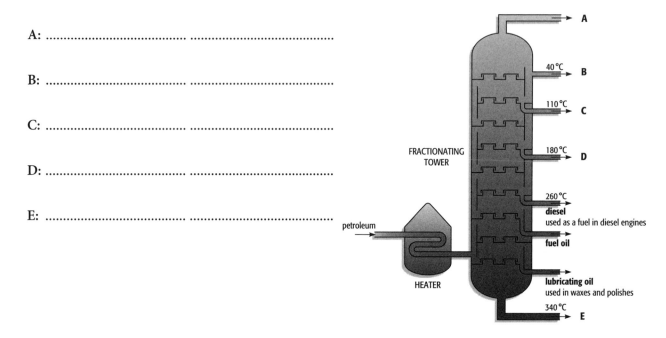

The table below shows the percentage by mass of some of these different fractions in petroleum. Also shown is the demand for each fraction expressed as a percentage.

Fraction	Number of carbon atoms per molecule	Percentage in petroleum/%	Percentage needed by the oil refinery to supply demand/%
A	1–4	4	11
B	5–9	11	22
C	10–14	12	20
D	14–20	18	15
waxes and E	over 20	23	4

b Which physical property is used to separate petroleum by fractional distillation?

....

c Define the term **cracking.**

....

....

d Use information from the table to explain how cracking helps an oil refinery match the supply of gasoline (petrol) with the demand for gasoline.

....

....

....

....

e The hydrocarbon $C_{15}H_{32}$ can be cracked to make propene and one other hydrocarbon.

i Write an equation for this reaction.

....

ii Draw the structure of propene.

Exercise 11.2 Addition polymerisation

This exercise will help you practise the representation of polymers and develop your understanding of their uses and the issues involved.

a Poly(ethene) is a major plastic used for making a wide variety of containers. Complete these sentences about poly(ethene) using words from this list.

acids   addition   condensation   ethane

polymerisation   ethene   monomers   polymer

Poly(ethene) is a .... formed by the ....

of .... molecules. In this reaction the starting molecules can be

described as ....; the process is known as ....

b Draw the structure of poly(ethene) showing at least two repeat units.

The structure below is that of an addition polymer.

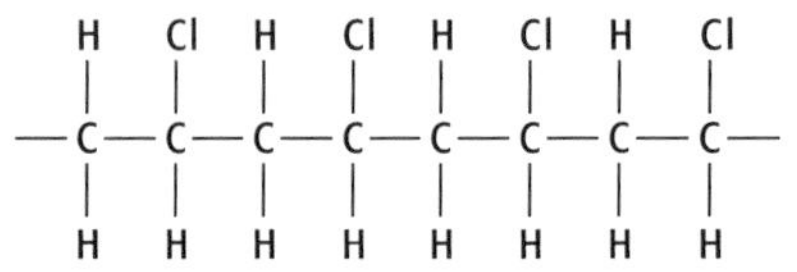

Polymer X

c Draw the structure of the monomer from which polymer **X** is formed.

d Polymer **X** is non-biodegradable. Describe one pollution problem that this causes.

....

....

e Polymer **X** can be disposed of by burning at high temperature. However, this can produce toxic waste gases such as hydrogen chloride. Hydrogen chloride can be removed from the waste gases by reaction with moist calcium carbonate powder. Name the three products of this reaction.

....

⑤ Exercise 11.3 The structure of man-made fibre molecules

This exercise is designed to help your understanding of the formation and structure of two man-made fibres.

These diagrams show sections of the polymer chain of two man-made polymers.

a Draw a circle around an amide link in the diagrams. Label this **amide link.**

Nylon

b Draw a circle around an ester link in the diagrams. Label this **ester link.**

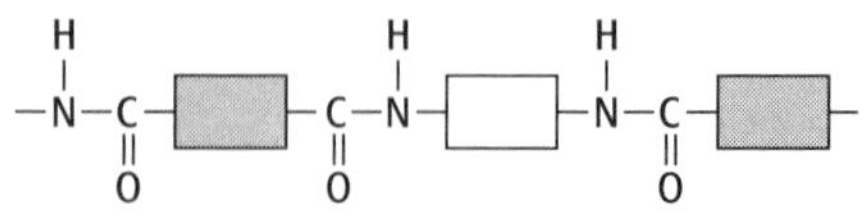

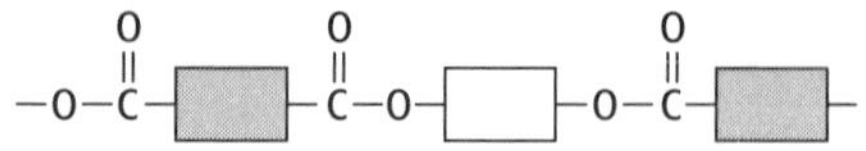

Terylene

c Name a type of naturally occurring polymer that has a similar link to nylon.

....

d The formulae of the two monomers used to make nylon are shown below.

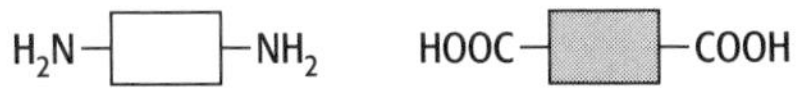

$H_2N-\boxed{}-NH_2$ $HOOC-\boxed{}-COOH$

Nylon monomers

Deduce the formulae of the two monomers that are used to make Terylene.

Terylene monomers

e State the functional groups on the monomers used to make Terylene.

....

f State the type of polymerisation that occurs when Terylene is made.

....

g State one large-scale use of Terylene.

....

h Name a naturally occurring class of compounds that contains the ester link.

....

Exercise 11.4 Condensation polymerisation

This exercise is aimed at developing your skills in extending and applying your knowledge from familiar material to unfamiliar examples.

Lactic acid polymerises to form the biodegradable polymer, polylactic acid (PLA). The monomer can be readily made from corn starch.

$CH_3-CH-COOH$
$|$
OH

Lactic acid

The structure of PLA is given here, showing the link joining the repeating units circled.

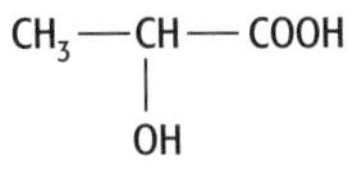

$$-O-\underset{CH_3}{\overset{}{CH}}-\underset{}{\overset{O}{C}}-O-\underset{CH_3}{\overset{}{CH}}-$$

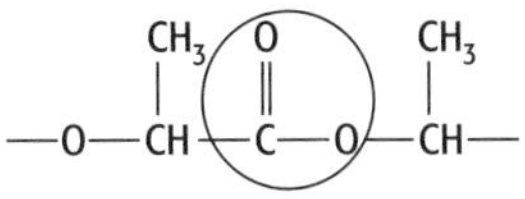

Polylactic acid (PLA)

5 a Suggest two advantages of PLA compared with a polymer made from petroleum.

....

....

b What type of compound contains the group that is circled?

....

c Complete the following sentences.

Lactic acid molecules can form this linking group because they contain two

different .... groups that react with each other to form

the .... link. Lactic acid molecules contain both

an .... group and an .... group.

d Is PLA formed by addition or condensation polymerisation? Give a reason for your choice.

....

....

When lactic acid is heated, acrylic acid is formed.

Lactic acid   Acrylic acid

e Complete the word equation for the action of heat on lactic acid.

lactic acid → .... + ...

f Describe a test that would distinguish between lactic acid and acrylic acid.

Test: ....

Result for lactic acid: ....

Result for acrylic acid: ....

⑨ Exercise 11.5 The analysis of condensation polymers

This exercise will develop your understanding of complex condensation polymers and their representation.

Enzymes are biological catalysts. Purified enzymes are used widely both in research laboratories and in industry.

Certain enzymes called proteases can hydrolyse proteins to amino acids. The amino acids can be separated and identified by chromatography. The diagram below shows a typical chromatogram.

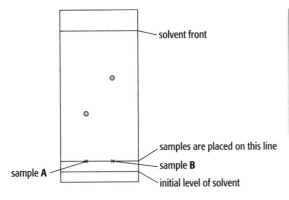

Amino acid	R_f value
leucine	0.9
alanine	0.7
glycine	0.5
glutamic acid	0.4

$$R_f \text{ value of a sample} = \frac{\text{distance moved by sample}}{\text{distance moved by solvent}}$$

a Using the table of R_f values for selected amino acids, identify the two amino acids on the chromatogram.

Sample A is ....

Sample B is ....

b Explain why the chromatogram must be exposed to a locating agent before the R_f values can be measured.

....

c Measuring R_f values is one way of identifying amino acids on a chromatogram; suggest another.

....

d Proteins can be hydrolysed chemically, without the use of enzymes. What are the conditions used for this hydrolysis?

....

e Compare the structure of a protein with that of a synthetic polyamide. The structure of a typical protein is given below.

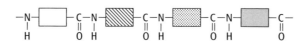

i How are they similar?

....

ii How are they different?

....

f A form of nylon can be made from the following two monomers.

H₂N—☐—NH₂   ClOC—▨—COCl

What would be the simple molecule released in the condensation reaction in this case?

....

Complex carbohydrates such as starch are another group of condensation polymers. Enzymes such as amylase, a carbohydrase, can hydrolyse complex carbohydrates to simple sugars which can be represented as

HO—☐—OH

g Draw the structure of the complex carbohydrate chain (showing at least three monomer units).

12 Chemical analysis and investigation

Useful equations

$AgNO_3(aq) + NaCl(aq) \rightarrow AgCl(s) + NaNO_3(aq)$ or $Ag^+(aq) + Cl^-(aq) \rightarrow AgCl(s)$

$BaCl_2(aq) + CuSO_4(aq) \rightarrow BaSO_4(s) + CuCl_2(aq)$ or $Ba^{2+}(aq) + SO_4^{2-}(aq) \rightarrow BaSO_4(s)$

$FeSO_4(aq) + 2NaOH(aq) \rightarrow Fe(OH)_2(s) + Na_2SO_4(aq)$ or $Fe^{2+}(aq) + 2OH^-(aq) \rightarrow Fe(OH)_2(s)$

Exercise 12.1 Titration analysis

This exercise will assist you in remembering some of the basic procedures involved in practical work, including points that will be checked on as you carry out coursework safely and rigorously.

a Choose words from the list below to complete the passage.

accurate	catalyst	phenolphthalein	indicator	measuring cylinder
pipette	qualitative	neutralised	quantitative	three

There are situations when chemists need to know how much of a substance is present,

or how concentrated a solution of a substance is. This type of experiment is part of

what is known as .... analysis. One experimental method used

here is titration.

The important pieces of apparatus used in titration are a burette and a

.... When an acid is titrated against an alkali, methyl

orange can be used as the .... so that we know that the acid has just

.... the alkali. A few drops of .... can be used as an alternative to methyl orange. The experiment is repeated several times, often

until .... results have been obtained that are in close agreement to each other.

b As part of an experiment to determine the value of x in the formula for iron(II) sulfate crystals ($FeSO_4.xH_2O$) a student titrated a solution of these crystals with $0.0200\,mol/dm^3$ potassium manganate(VII) (solution **A**).

A $25.0\,cm^3$ sample of the iron(II) sulfate solution was measured into a conical titration flask. Solution **A** was run from a burette into the flask until an end-point was reached. Four titrations were done. The diagrams show parts of the burette before and after each titration.

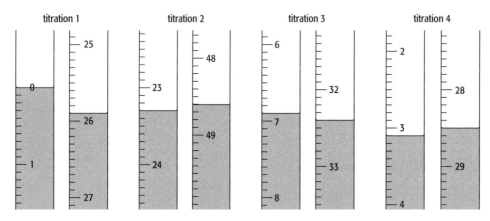

i Use the diagrams to complete the table of results.

Titration number	1	2	3	4
Final burette reading / cm³	....	....	....	....
First burette reading / cm³	....	....	....	....
Volume of solution A / cm³	....	....	....	....
Best titration results (✓)				

Tick (✓) the columns with the best titration results. Using these results, the average

volume of A was .... cm³.

ii Solution **A** is $0.0200\,mol/dm^3$ potassium manganate(VII). Calculate how many moles of $KMnO_4$ were present in the titrated volume of **A** calculated in part **i**.

iii What type of reaction takes place between the iron(II) sulfate solution and solution **A**?

....

iv Potassium manganate(VII) is purple. What was the colour change at the end-point?

Change from .... to ....

Exercise 12.2 Chemical analysis

This exercise will help familiarise you with some of the analytical tests and the strategy behind them. Remember that these tests can come up frequently on the written papers as well as the practical papers.

a The following table shows the tests that some students did on substance **A** and the conclusions they made from their observations.

i Complete the table by describing these observations and suggest the test and observation which led to the conclusion in test **4**.

Test	Observation	Conclusion
1 Solid **A** was dissolved in water and the solution divided into three parts for tests **2**, **3** and **4**.	....	A does not contain a transition metal.
2 (i) To the first part, aqueous sodium hydroxide was added until a change was seen. (ii) Excess aqueous sodium hydroxide was added to the mixture from (i).	....	A may contain Zn^{2+} ions or Al^{3+} ions.
3 (i) To the second part aqueous ammonia was added until a change was seen. (ii) An excess of aqueous ammonia was added to the mixture from (i).	....	The presence of Zn^{2+} ions is confirmed in A.
4	....	A contains I^- ions.

ii Give the name and formula of compound **A**.

....

b A mixture of powdered crystals contains both ammonium ions (NH_4^+) and zinc ions (Zn^{2+}). The two salts contain the same anion (negative ion). The table below shows the results of tests carried out by a student.

i Complete the table of observations made by the student.

Test	Observations
1 A sample of the solid mixture was dissolved in distilled water. The solution was acidified with dilute HCl(aq) and a solution of $BaCl_2$ added.	A white precipitate was formed.
2 A sample of the solid was placed in a test tube. NaOH(aq) was added and the mixture warmed. A piece of moist red litmus paper was held at the mouth of the tube.	The solid dissolved and pungent fumes were given off. The litmus paper turned ...., indicating the presence of .... ions.
3 A sample of the solid was dissolved in distilled water to give a .... solution. NaOH(aq) was added dropwise until in excess.	A .... precipitate was formed which was .... in excess alkali.
4 A further sample of the solid was dissolved in distilled water. Concentrated ammonia solution (NH_3(aq)) was added dropwise until in excess.	A .... precipitate was formed. On addition of excess alkali the precipitate was ....

ii Give the names and formulae of the two salts in the mixture.

....

iii Give the name and formula of the precipitate formed in tests **3** and **4**.

....

Exercise 12.3 Experimental design

This exercise emphasises the considerations that are important when planning and evaluating an experimental method.

How is the rate of a reaction affected by temperature?

The reaction between dilute hydrochloric acid and sodium thiosulfate solution produces a fine yellow precipitate that clouds the solution. This means that the rate of this reaction can be found by measuring the time taken for a cross (X) under the reaction to become hidden.

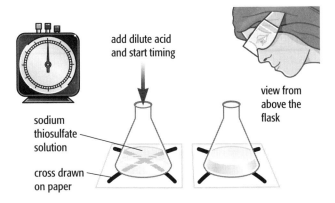

a   You are asked to design an experiment to see how changing the temperature of the solutions mixed affects the rate of the reaction.
   You are provided with the following apparatus and solutions:

- several 100 cm³ conical flasks of the same size and shape
- dilute hydrochloric acid solution (0.5 moles per dm³)
- sodium thiosulfate solution – a colourless solution (0.5 moles per dm³)
- several 50 cm³ measuring cylinders
- a piece of white card and a felt tip marker pen
- a stopclock
- a water bath that is thermostatically controlled so that the temperature can be adjusted – flasks of solution can be placed in this to adjust to the required temperature
- two thermometers
- and any other normal lab apparatus.

Your description should include:

- a statement of the **aim** of the experiment – comment on which factors in the experiment need to be kept constant and why
- a description of the **method** for carrying out the experiment – this should be a **list of instructions** to another student
- safety – put in a comment on what **safety precautions** you need to take and why.

....

....

....

....

....

....

....

....

b Below are the results of tests carried out at five different temperatures. In each case 50 cm^3 of aqueous sodium thiosulfate was poured into a flask. 10 cm^3 of hydrochloric acid was added to the flask. The initial and final temperatures were measured. Use the thermometer diagrams to record all of the initial and final temperatures in the table.

i Complete the table of results to show the average temperatures.

Experiment	Thermometer diagram at start	Initial temperature /°C	Thermometer diagram at end	Final temperature /°C	Average temperature /°C	Time for cross to disappear /s
1	30 / 25 / 20	....	30 / 25 / 20	....	....	130
2	40 / 35 / 30	....	40 / 35 / 30	....	....	79
3	45 / 40 / 35	....	45 / 40 / 35	....	....	55
4	55 / 50 / 45	....	55 / 50 / 45	....	....	33
5	60 / 55 / 50	....	60 / 55 / 50	....	....	26

ii Plot a graph of the time taken for the cross to disappear versus the average temperature on the grid and draw a smooth line graph.
Use the check list below to give yourself a mark for your graph.

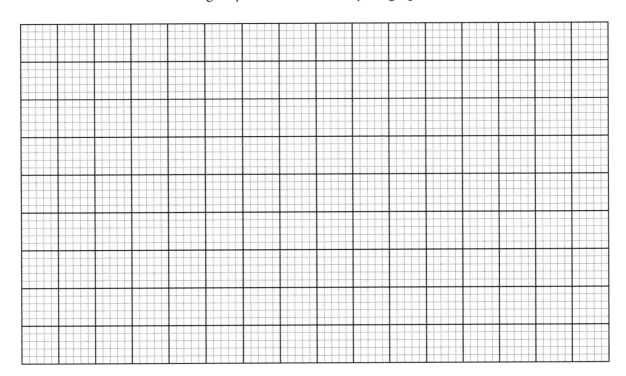

For each point, award yourself: 2 marks if you did it really well
1 mark if you made a good attempt at it, and partly succeeded
0 marks if you did not try to do it, or did not succeed.

Self-assessment check list for a graph:

Check point	Marks awarded	
	You	Your teacher
You have drawn the axes with a ruler, using most of the width and height of the grid.		
You have used a good scale for the x-axis and the y-axis, going up in useful proportions.		
You have labelled the axes correctly, giving the correct units for the scales on both axes.		
You have plotted each point precisely and correctly.		
You have used a small, neat dot or cross for each point.		

Check point	Marks awarded	
	You	Your teacher
You have drawn a single, clear best-fit line through the points – using a ruler for a straight line.		
You have ignored any anomalous results when drawing the line.		
Total (out of 14)		

12–14 Excellent.
10–11 Good.
7–9   A good start, but you need to improve quite a bit.
5–6   Poor. Try this same graph again, using a new sheet of graph paper.
1–4   Very poor. Read through all the criteria again, and then try the same graph again.

c In which experiment was the speed of reaction greatest? ....

d Explain why the speed was greatest in this experiment.

....

....

e Why were the same volume of sodium thiosulfate solution and the same volume of hydrochloric acid used in each experiment? Why do the conical flasks used in each test run need to be of the same dimensions?

....

....

f From the graph, deduce the time for the cross to disappear if the experiment was to be repeated at 70 °C. Show clearly on the grid how you worked out your answer.

....

g Sketch on the grid the curve you would expect if all the experiments were repeated using 50 cm^3 of more concentrated sodium thiosulfate solution.

h How would it be possible to achieve a temperature of around 0 to 5 °C?

....

i Explain one change that could be made to the experimental method to obtain more accurate results.

....

....